ON OUR TERMS

U.S. MARINES IN OPERATION DEWEY CANYON
22 JANUARY TO 18 MARCH 1969

SETH A. GIVENS, PHD

MARINES IN THE VIETNAM WAR COMMEMORATIVE SERIES

This pamphlet history, one of a series devoted to U.S. Marines in the Vietnam War, is published for the education and training of Marines by the History Division, Marine Corps University, Quantico, Virginia, as part of the U.S. Department of Defense fiftieth anniversary of that war. Editorial costs have been defrayed in part by contributions from members of the Marine Corps Heritage Foundation.

LEGAL NOTICE

Marine Corps History Division is the service history office of the United States Marine Corps. As a product of the U.S. government, no copyright protection is attached. However, authors, contributors, and designers may retain intellectual property rights to their work product. Users should check with the appropriate office of the U.S. government before altering and/or reproducing this material to ensure that such work does not violate intellectual property right protections. The information contained in this publication was deemed accurate at the time of printing. Comments and/or corrections should be sent to the Marine Corps History Division for consideration in future printings.

The production of this work and other History Division products is graciously supported by the Marine Corps Heritage Foundation.

THIS VOLUME IS FREELY AVAILABLE AT WWW.USMCU.EDU/HDPUBLISHING

2021

Library of Congress Control Number: 2021931337

INTRODUCTION

At Vandegrift Combat Base, the home of 9th Marines, Brigadier General Frank E. Garretson assembled his staff on 14 January 1969. As commanding general of Task Force Hotel, he was responsible for 3d Marine Division's area of operations in the northwestern reaches of the I Corps Tactical Zone, the northernmost of four political and military regions in the Republic of Vietnam.[1] The next day, the Task Force Hotel commander sent a message to 9th Marines and 2d Battalion, 12th Marines, both of which were southwest of Vandegrift Combat Base searching for any sizable enemy concentration near Khe Sanh.[2] He ordered his infantry and artillery commanders to plan immediately for a regiment-size search and clear operation to their southeast, where Laos jaggedly protrudes into the Republic of Vietnam's Quang Tri Province, in the rugged and lush Da Krong Valley.[3]

The orders originated from 3d Marine Division headquarters at Dong Ha, 25 kilometers northeast of Vandegrift Combat Base, where Major General Raymond G. Davis and his staff monitored increased activity from North Vietnamese Army (NVA) units, officially known as the People's Army of Vietnam.[4] The first indication was a communication wire running along a trail system.[5] Enemy engineer units that were inactive for several months now reopened Routes 922 and 548, branches of the Ho Chi Minh Trail that serviced a logistics depot in Laos and the Republic of Vietnam named Base Area 611. Helicopters and reconnaissance aircraft even took fire from 12.7mm, 25mm, and 37mm antiaircraft weapons where they never had before, leading to the downing of a U.S. Navy Grumman A-6 Intruder all-weather attack air-

---

[1] The four tactical zones in the Republic of Vietnam were, north to south, I, II, III, and IV Corps. I Corps was pronounced "Eye" Corps. Throughout this volume, the Republic of Vietnam refers to South Vietnam and the Democratic Republic of Vietnam refers to North Vietnam.

[2] Col Robert H. Barrow tto MajGen Raymond G. Davis, "Combat Operation After Action Report," 8 April 1969, box 9, folder 14, Vietnam War Collection, Archives Branch, Marine Corps History Division, Quantico, VA (hereafter HD Archive). "Regiment" often appears in the title of Marine units—e.g., 9th Marine Regiment—but the Marine Corps does not officially recognize its usage. The Service first used "Marines" in reference to regiments during World War I, when "Regiment of Marines" was the preferred term to differentiate the Marine units placed under command of the U.S. Army. In 1930, the Marine Corps dropped "Regiment" from official designations and merely referred to both infantry and artillery regiments as the number designator plus "Marines." That naming convention remains today.

[3] Col Robert H. Barrow, interview with SSgt Willis S. Bernard Jr., 8 April 1969, Marine Corps History Division Oral History Collection, Archives Branch, Marine Corps History Division, Quantico, VA (hereafter Marine Corps Oral History Collection), hereafter Barrow interview.

[4] American personnel used the term North Vietnamese Army when discussing regular forces from the Democratic Republic of Vietnam. In keeping with this tradition, this pamphlet will primarily use NVA, substituting People's Army of Vietnam where it is appropriate. For message traffic, see G-2 and G-3 Journal, 3d Marine Division, January 1969, HD Archive.

[5] Gen Raymond G. Davis, interview with Benis M. Frank, 2 February 1977, transcript, Marine Corps Oral History Collection, hereafter Davis interview.

craft.[6] By early January, 1,000 trucks rolled down the Ho Chi Minh Trail every day. American aircraft attempted to close the important infiltration routes that supplied staging areas in the Republic of Vietnam, but enemy engineers repaired the damage and trucks continued delivering supplies. Both supply routes were vital to operations that Hanoi aimed at destabilizing the government in Saigon, as the east-west Route 922 gave NVA troops the ability to move from Laos into the Republic of Vietnam and the strategically important A Shau Valley via Route 548.[7]

For years, NVA forces used the A Shau Valley as an infiltration corridor into the south. Troops staged weapons and ammunition caches to supply the Viet Cong, a guerrilla organization in the Republic of Vietnam and the military arm of the National Liberation Front.[8] Allied with Hanoi, the National Liberation Front was a Communist revolutionary organization that aimed to reunite the two halves of Vietnam under one government. The Viet Cong used the weapons to launch attacks throughout the countryside and important urban areas like Hue and Da Nang.

Terrain in the two valleys made it difficult for American and Republic of Vietnam units to operate. Tall elephant grass in the Da Krong Valley's kilometer-wide plain obscured enemy units and ridgelines that reached 1,500 meters with 20–45-degree slopes hindered maneuver. Unpredictable weather and low cloud cover added to the complications, as both cut off the A Shau and Da Krong Valleys from the rest of I Corps during the monsoon season. NVA units used the local conditions to their advantage and established sanctuaries in the remote areas. During Operation Delaware in April and May 1968, elements of the U.S. Army's 1st Cavalry Division (Airmobile), 101st Airborne Division, and the Army of the Republic of Vietnam's (ARVN) Airborne Task Force attempted to exploit battlefield gains after the Tet Offensive. They hoped to eliminate the enemy's important staging area in the A Shau Valley to forestall any future attacks in the I and II Corps Tactical Zones. Three months later, 1st Brigade, 101st Airborne Division, returned with the 1st ARVN Division to follow up with Operation Somerset Plain.[9] Since then, the area had been quiet. Now, it appeared NVA units were sending supplies from Laos into the A Shau Valley for another winter-spring offensive during the Tet holiday.

To prevent enemy attacks along the coastal plain on American and ARVN forces, government authorities, and public infrastructure, the Marines planned a regimental attack on NVA sanctuaries in the mountains of Quang Tri Province. Codenamed Operation Dewey Canyon, the preemptive action was significant because of the objective and the way in which the Marines executed it. An adaptation of the U.S. Army's airmobility concept, what Major General Davis termed *high mobility* was 3d Marine Division's rebuke of the defensive posture units took in I Corps since 1967. No longer would battalions remain in fixed positions waiting to interdict enemy units. The 3d Marine Division, as Davis put it, was not going to "sit there and absorb the shot and shell." Its regiments would instead be mobile, aggressive, and decisive. Operation Dewey Canyon, the last large-scale Marine offensive of the Vietnam War, would demonstrate that the Marines were resolved to destroy the enemy "on our terms."[10]

To understand the operation, it is important to understand the concept behind it. This commemorative volume is about high mobility and focuses on Operation Dewey Canyon as a well-known example. *High mobility* was an air assault concept that relied on helicopters to insert infantry-artillery teams into enemy-controlled areas. It emphasized projecting combat power via short-term, mountaintop fire support bases from which artillery supported infantry sweeps of territory. High mobility was a tactical refinement to achieve an operational end. It was also the product of learning, built on Marine

---

[6] MajGen Raymond G. Davis to LtGen Robert E. Cushman Jr., "Artillery Report of Operation Dewey Canyon," 5 May 1969, box 9, folder 19, Vietnam War Collection, HD Archive.

[7] U.S. Air Force, "Resume of the A Shau Valley—9 December 1968–28 February 1969," box 9, folder 85, Contemporary Historical Examination of Current Operations (CHECO) Reports of Southeast Asia, 1961–1975, Vietnam Center and Sam Johnson Vietnam Archive, Texas Tech University, Lubbock, TX (hereafter TTU Vietnam Archive).

[8] The term *Viet Cong* is a colloquialism meaning "Vietnamese Communist" that originated in the 1920s to differentiate the group from Chinese Communists. While some have assigned pejorative value to the term, it is used here purely as a descriptive identifier. The organization is also known as the Liberation Army of South Vietnam and the People's Liberation Armed Forces (not to be confused with the People's Republic of China's armed forces, known as the People's Liberation Army). See Brett Reilly, "The True Origin of the Term 'Viet Cong'," *Diplomat*, 31 January 2018.

[9] LtGen John J. Tolson, *Airmobilty, 1961–1971*, Vietnam Studies (Washington, DC: Department of the Army, 1999), 182–92. See also Shelby L. Stanton, *Anatomy of a Division: The 1st Cav in Vietnam* (Novato, CA: Presidio Press, 1987), 143.

[10] Davis interview.

Marines on patrol outside Chu Lai.
John T. Dyer Collection, Archives Branch, Marine Corps History Division

and Army units' experiences defending the Republic of Vietnam against NVA forces between March 1965 and fall 1968.

STRATEGIC AND OPERATIONAL BACKGROUND

Marines were confined to a limited mission of airfield security when the first battalion of 9th Marine Expeditionary Brigade (9th MEB) landed at Red Beach near Da Nang on 8 March 1965. Throughout 1965, Marine involvement in I Corps evolved from a limited defensive mission of protecting installations along the coastline to a balance between installation defense and rolling back the National Liberation Front's control of villages. The shift to a mixture of defensive and offensive stances led to 9th MEB's expansion into III Marine Amphibious Force (III MAF), with Major General Lewis W. Walt in command. The newly formed headquarters extended its troops' presence north and south of Da Nang to two other enclaves: Phu Bai and Chu Lai.[11] III MAF believed a balanced strategy of search and destroy, counterguerrilla, and pacification operations had five tasks: the defense and development of base areas, support of ARVN operations, offensive operations against the National Liberation Front, support of contingencies outside of I Corps, and the pursuit of pacification through gaining the support of the Republic of Vietnam's rural population.[12]

The three enclaves of Da Nang, Phu Bai, and Chu Lai were the focus of Marine operations in Vietnam. In Major General Walt's estimate, the extra troops and three-pronged approach could join the enclaves, thus securing the 400 kilometers of I Corps' coastal plain by the end of 1966.[13] By coordinating with local authorities and security forces, III MAF believed it possible to push guerrillas from the villages in the immediate vicinity of the enclaves and then spread security, as the logic went, like an inkblot.[14] Their assumption, however, was not built solely on a belief in projecting force. The pacification campaign the Marines aimed at dismantling Communist infrastructure in villages included civic action programs that offered medical assistance, community development, and security assurances from Combined Action Platoons—Popular Forces militias with Marine squads attached.[15] Despite the expansion of the Marines' tactical area of responsibility and their interaction with local populations, leaders like Walt still believed that winning hearts and minds was the purview of the U.S. State Department. III MAF's primary focus remained security, which Walt believed subordinate units could achieve through the combination of search and destroy, counterguerrilla, and pacification operations.[16]

The Marine approach to pacification offered an alternative to the emphasis at U.S. Military Assistance Command, Vietnam (USMACV), where Army General William C. Westmoreland stressed seeking out and destroying Viet Cong units.[17] Pacification was not a binary choice between clear and hold

---

[11] Jack Shulimson and Maj Edward F. Wells, "First In, First Out: The Marine Experience in Vietnam, 1965–71," in *The Marines in Vietnam, 1954–1973: An Anthology and Annotated Bibliography*, 2d ed. (Washington, DC: History and Museums Division, Headquarters, U.S. Marine Corps, 1985), 26; and BGen Edwin H. Simmons, "Marine Corps Operations in Vietnam, 1965–1966," in *The Marines in Vietnam, 1954–1973*, 42–46.

[12] Simmons, "Marine Corps Operations in Vietnam, 1965–1966," 56.
[13] Shulimson and Wells, "First In, First Out," 27.
[14] David Strachan-Morris, *Spreading Ink Blots from Da Nang to the DMZ: The Origins and Implementation of U.S. Marine Corps Counterinsurgency Strategy in Vietnam, March 1965 to November 1968* (Warwick, UK: Helion and Company, 2020), 86–115.
[15] MSgt Ronald E. Hays, USMC (Ret), *Combined Action: U.S. Marines Fighting A Different War, August 1965 to September 1970*, Marines in the Vietnam War Commemorative Series (Quantico, VA: Marine Corps History Division, 2019).
[16] Michael A. Hennessy, *Strategy in Vietnam: The Marines and Revolutionary Warfare in I Corps, 1965–1972* (Westport, CT: Praeger, 1997), 69.
[17] Andrew F. Krepinevich Jr., *The Army and Vietnam* (Baltimore, MD: Johns Hopkins University Press, 1986), 194–214.

I CORPS TACTICAL ZONE

Map courtesy of Pete McPhail, adapted by MCUP

Kim Le Bat, village chief of Thuy Phy, points out an enemy artillery concentration area to MajGen Lewis W. Walt (wearing helmet) and LtCol William W. Taylor (far left), commanding officer of 3d Battalion, 4th Marines.

or search and destroy, however.[18] The former could support the latter.[19] Westmoreland's instructions to Walt at the end of 1965 were to defend the Marine base areas, seek out and destroy Viet Cong guerrillas that posed a threat, and execute contingency plans wherever USMACV directed.[20] The Marine concept was perhaps more nuanced than the U.S. Army's, but the two could coexist and III MAF could still execute its tasks of waging conventional and counterinsurgency warfare.[21]

That coexistence had much to do with the peculiarities of I Corps. With 16,000 square kilometers that was both military zone and political region, I Corps was larger than Connecticut and made up one-sixth of the total size of the Republic of Vietnam. When Marines began ranging from Da Nang, Phu Bai, and Chu Lai in spring 1965, they possessed little ability to cover I Corps' five provinces (Quang Tri, Thua Thien, Quang Nam, Quang Tin, and Quang Ngai) and 2.6 million people.

---

[18] For an alternative view, see John A. Nagl, *Learning to Eat Soup with a Knife: Counterinsurgency Lessons from Malaya and Vietnam* (Chicago, IL: University of Chicago Press, 2005); and Lewis Sorley, *Westmoreland: The General Who Lost Vietnam* (New York: Houghton Mifflin, 2011).
[19] Gregory A. Daddis, *Westmoreland's War: Reassessing American Strategy in Vietnam* (New York: Oxford University Press, 2014), 105–9.
[20] Shulimson and Wells, "First In, First Out," 27.

[21] Nicholas J. Schlosser, "Reassessing the Marine Corps' Approach to Strategy in the Vietnam War, 1965–1968," *International Bibliography of Military History* 34, no. 1 (June 2014): 27–52, https://doi.org/10.1163/22115757-03401005.

The bulk of the population centers were in the east, along the coastal beaches and fertile deltas that extend deep into the interior. I Corps' borders were, for the most part, natural. The demilitarized zone, which divided the Democratic Republic of Vietnam from the Republic of Vietnam, used the rough outline of the Ben Hai River as the demarcation. In the west, the 1,000-kilometer-long Annamite Range made much of the western half of I Corps difficult to operate in and sparsely populated. More than 350 kilometers south of the demilitarized zone, the spur of the Annamites that juts toward the coast formed the southern boundary of I Corps. Lines of communication were scarce and therefore vital. Route 1 serviced the length of the Republic of Vietnam, from the demilitarized zone to Saigon, as the primary coastal road. Emanating from Route 1 and heading the varying 50–150 kilometers west through the river valleys and mountains to the Laotian border were Routes 9 and 14, in the north and south of I Corps, respectively. If the Marines controlled those critical roads, they could link together the population centers and supply units in the more rugged areas.[22]

OPERATIONS INTO WESTERN QUANG TRI PROVINCE

The character of the war in I Corps changed in 1966. III MAF launched operations to destroy Viet Cong guerrillas and, in some cases, NVA units that entered Marine tactical areas of responsibility. While Combined Action Platoons operated on their own, rifle companies and infantry battalions conducted pacification operations. Some units enacted civic action lessons from the Marine Corps' 1940 *Small Wars Manual*, but Marine emphasis remained on overwhelming force: units provided security to villages via cordons and searches to separate guerrillas from local populations and set the conditions for government programs.[23] Most unit engagements early in the year were against Viet Cong guerrillas in the lowlands near Hue, Da Nang, and Chu Lai. As Marines gradually expanded into the hills and jungles of northwestern I Corps, they engaged NVA units. In July 1966, Operation Hastings initiated a multiple-battalion clash with regular troops infiltrating Quang Tri Province. The fighting near the 200-meter-tall feature known as the

Douglas A. Yeager Collection, Archives Branch, Marine Corps History Division

The rugged terrain of northern Quang Tri Province that the Marines went into in 1966. The massif in the center of the valley is the Rockpile.

Rockpile, 40 kilometers west of Dong Ha, was important not only for deciding who controlled Route 9 but also because it signaled a Marine push into the mountainous terrain of central and western Quang Tri Province. Battalions returned the next month in Operation Prairie, which lasted into January 1967.[24]

By summer 1966, it became clear to III MAF that Marine battalions would remain in the rugged areas of northern I Corps, where the enemy was determined to fight. As a consequence, III MAF conducted pacification operations in the coastal lowlands of I Corps with 1st Marine Division while 3d Marine Division waged a conventional campaign in the hills between Route 9 and the demilitarized zone.[25] III MAF's emphasis at the end of 1966 was on creating breathing space for pacification to take hold in the coastal areas.[26] In October, Westmoreland and Walt ordered 3d Marine Division to west-

---

[22] Simmons, "Marine Corps Operations in Vietnam, 1965–1966," 40–42.
[23] Dr. Nicholas J. Schlosser, "Marine Corps' Small Wars Manual: An Old Solution to a New Challenge?," *Fortitudine* 35, no. 1 (2010): 9.

[24] Jack Shulimson, *U.S. Marines in Vietnam: An Expanding War, 1966* (Washington, DC: History and Museums Division, Headquarters, U.S. Marine Corps, 1982), 231–43.
[25] Simmons, "Marine Corps Operations in Vietnam, 1965–1966," 56; and Shulimson and Wells, "First In, First Out," 27.
[26] Hennessy, *Strategy in Vietnam*, 78–79.

3d Battalion, 4th Marines, position overlooking the western approaches to Phu Bai, October 1965.

ern Quang Tri Province to serve as a blocking force against NVA soldiers attempting to supply and support Viet Cong guerrillas farther inside the Republic of Vietnam.[27] As Westmoreland saw it, USMACV's strategy was akin to a boxer jabbing at his opponent (enemy main forces) while protecting his body (the population).[28] As a result, the conventional and pacification operations in I Corps were distinct but connected: beat back NVA troops in the mountains with conventional means so that units in the lowlands could destroy the isolated Viet Cong guerrillas and secure pacification.[29]

The Democratic Republic of Vietnam's commander in chief of the armed forces and minister of defense, Vo Nguyen Giap, confirmed the logic behind III MAF's operational plan later in the year in a series of articles published in the armed forces newspaper *Quan Doi Nhan Dan*. "The Marines," Giap claimed, "are being stretched as taut as a bowstring over hundreds of kilometers."[30] Giap confirmed what Walt and Lieutenant General Victor H. Krulak, commanding officer of Fleet Marine Force, Pacific, believed—that Hanoi's strategy was to undermine pacification efforts in the lowlands by coaxing American units away from villages and towns to the border areas.[31] In effect, both sides were inducing each other to behave as they wanted, fighting conventionally in the northern stretches of I Corps to win the battle over the populated areas nearer the coastline. This placed considerable significance on the fight near the demilitarized zone.

Giap displayed his strategy in Quang Tri Province in the first half of 1967, when 3d Marine Division, busy constructing firebases along Route 9, repelled significant North Vietnamese attacks near Con Thien.[32] The first pitched battle near the former Special Forces camp at Khe Sanh occurred in April and served as a prologue to a summer full of skirmishes with enemy units attempting to cut the vital line of communication that supplied reinforcements to the firebases.[33] In the eastern half of the province near the demilitarized zone, the Marines conducted several battalion-size operations, among them Operations Cimarron and Buffalo, to reduce the enemy pressure on Con Thien. Fighting south of the demilitarized zone between Con Thien, Gio Linh, Dong Ha, and Cam Lo, an area that Marines dubbed "Leatherneck Square," became frenetic at times. For Giap, the repeated attacks on the enclaves turned Con Thien and Da Nang into "isolated islands in the open sea of people's war."[34] His comment was accurate to the extent that USMACV and Washington were thinking of ways to protect the islands from NVA infiltration. The American solution was, in keeping with Giap's metaphor, to create an ocean that the enemy could not cross.

THE RISE AND FALL OF THE STRONGPOINT DEFENSE

In September 1966, Secretary of Defense Robert S. McNamara announced the plan for an anti-infiltration barrier the length of the demilitarized zone, from the South China Sea to the Laotian border.[35] The strongpoint defense system, which some referred to as the McNamara Line, was unmanned, patrolled from the air, and featured a combination of minefields and trenches with seismic and motion detectors.[36] General Westmoreland doubted the system could work, and he proposed to Secretary McNamara a compromise barrier and strongpoint system instead. In the east, USMACV could use a manned version of the deforested barrier system with an armored cavalry regiment as a mobile force to patrol and defend the trace. In the west, where the hills of Quang Tri Province made a barrier system difficult, Westmoreland advo-

---

[27] Shulimson, *An Expanding War, 1966*, 314.
[28] Graham A. Cosmas, *MACV: The Joint Command in the Years of Escalation, 1962–1967* (Washington, DC: Center of Military History, United States Army, 2006), 403–4.
[29] Edward Thomas Nevgloski, "Understanding the United States Marines' Strategy and Approach to the Conventional War in South Vietnam's Northern Provinces, March 1965–December 1967" (PhD diss., King's College London, 2019).
[30] Vo Nguyen Giap, "The Big Victory, the Great Task," *Quan Doi Nhan Dan*, 14–16 September 1967.
[31] See Victor H. Krulak, *First to Fight: An Inside View of the U.S. Marine Corps* (Annapolis, MD: Naval Institute Press, 1984), 179–204.
[32] Col Joseph C. Long, USMCR (Ret), *Hill of Angels: U.S. Marines and the Battle for Con Thien, 1967 to 1968*, Marines in the Vietnam War Commemorative Series (Quantico, VA: Marine Corps History Division, 2016).
[33] Col Rod Andrew Jr., USMCR, *Hill Fights: The First Battle of Khe Sanh, 1967*, Marines in the Vietnam War Commemorative Series (Quantico, VA: Marine Corps History Division, 2017).
[34] DIA Intelligence Supplement, "General Vo Nguyen Giap's Present View of the War," 25 September 1967, box 13, folder 18, Glenn Helm Collection, TTU Vietnam Archive.
[35] Edward J. Drea, *McNamara, Clifford, and the Burdens of Vietnam, 1965–1969*, vol. VI, Secretaries of Defense Historical Series (Washington, DC: Historical Office, Office of the Secretary of Defense, 2011), 127–30.
[36] LtGen Willard Pearson, *The War in the Northern Provinces, 1966–1968*, Vietnam Studies (Washington, DC: Department of the Army, 1991), 21–24.

Douglas A. Yeager Collection, Archives Branch, Marine Corps History Division

Marine combat base at Con Thien, a position only a few kilometers from the Democratic Republic of Vietnam and the northwest corner of Leatherneck Square.

cated for a strongpoint defense of mutually supporting firebases.[37]

Walt and his subordinates believed strongpoints were impractical because they tied troops to a barrier. A division-size mobile force, they argued, was a more flexible and economical approach to blocking enemy infiltration. McNamara's concept, they believed, was more a counter to guerrilla warfare than the conventional war along the demilitarized zone. After all, it was regular soldiers coming into the Republic of Vietnam, not irregular fighters. Most bothersome to Walt, the strongpoint defense violated his concept of the war in I Corps. By emphasizing the infiltration problem, USMACV forgot that the struggle over the population centers in the coastal plain was the real fight and the National Liberation Front was the primary threat.[38]

Construction of the firebreak was slow due to a lack of resources and manpower and vacillating interest from USMACV.[39] In June 1967, Lieutenant General Robert E. Cushman Jr. replaced Walt as III MAF commander. Unlike his predecessor, Lieutenant General Cushman supported the anti-infiltration system, believing it allowed him to move units where he needed them most. Construction still lagged, leading Cushman to request more troops from USMACV in August to finish the trace before the monsoon season began in November. At this point, Westmoreland's support for the anti-infiltration barrier eroded. Enemy attacks continued while the Marines worked on the trace. Lacking sufficient manpower to do both simultaneously, Westmoreland ordered III MAF to stop constructing the barrier until the Marines could stabilize the tactical situation. In western Quang Tri Province, however, 3d Marine Division was to continue building the strongpoint defense.[40]

To reinforce the battalions stretched thin across I Corps, Westmoreland ordered Operation Checkers in late December, a reshuffling of units farther north.[41] Cushman moved all of 3d Marine Division to the demilitarized zone and Khe Sanh area, placing its command post forward, from Phu Bai to Dong Ha. To fill the vacuum, 1st Marine Division inherited the Phu Bai enclave as part of its tactical area of responsibility.[42] While Operation Checkers was ongoing, Westmoreland, who worried about an enemy campaign that would sweep over Quang Tri and Thua Thien Provinces, sent north his last reserves, the U.S. Army's 1st Cavalry Division (Airmobile) and 2d Brigade, 101st Airborne Division.[43] They joined two other Army units already in I Corps, the 23d Infantry Division and 3d Brigade, 1st Cavalry Division (Airmobile).[44] All reinforcements fell under operational control of III MAF, giving Cushman a brigade, two Army divisions, two Marine divisions, an aircraft wing, and a host of supporting forces.[45] Westmoreland ordered

---

[37] Shulimson and Wells, "First In, First Out," 27–28.
[38] Shulimson, *An Expanding War, 1966*, 318–19.

[39] Simmons, "Marine Corps Operations in Vietnam, 1967," in *The Marines in Vietnam, 1954–1973*, 86–87; and Shulimson and Wells, "First In, First Out," 28.
[40] Shulimson and Wells, "First In, First Out," 28.
[41] Jack Shulimson et al., *U.S. Marines in Vietnam: The Defining Year, 1968* (Washington, DC: History and Museums Division, Headquarters, U.S. Marine Corps, 1997), 83.
[42] Shulimson and Wells, "First In, First Out," 29.
[43] Cosmas, *MACV*, 72–74.
[44] The 23d Infantry Division is better known as the Americal Division.
[45] Simmons, "Marine Corps Operations in Vietnam, 1967," 102.

Commandant of the Marine Corps Gen Wallace M. Greene Jr. (left), LtGen Robert E. Cushman Jr. (center), and Gen William C. Westmoreland (right) meet at III MAF headquarters, 10 August 1967.

the reinforcements because of conditions as well as intelligence that sizable enemy units were on the move. By mid-January 1968, he confirmed that the forces were heading for Saigon, Hue, Da Nang, Khe Sanh, the demilitarized zone, and several provincial capitals.[46]

Westmoreland and Cushman both saw the Marines at Khe Sanh as a vulnerable but vital blocking force. They feared North Vietnamese troops occupying the Khe Sanh area would give the enemy a clear invasion route into Quang Tri and Thua Thien Provinces and the opportunity to outflank III MAF forces in the border area. With little choice but to stand and fight, Cushman prepared to hold the Marines' vital lines of communication.[47]

Between 29 and 31 January 1968, enemy forces attacked I Corps' major population centers as part of the Tet Offensive. The NVA general offensive and National Liberation Front's popular uprising, though startling in size and ferocity, were both tactical failures and political victories.[48] American and ARVN units defeated the enemy on the battlefield to such an extent that the Viet Cong struggled for the next two years

---

[46] *Report on the War in Vietnam* (Washington, DC: Department of Defense, 1968), 157; and BGen Edwin H. Simmons, "Marine Corps Operations in Vietnam, 1968," in *The Marines in Vietnam, 1954–1973*, 102.

[47] Shulimson and Wells, "First In, First Out," 29.
[48] David F. Schmitz, *The Tet Offensive: Politics, War, and Public Opinion* (Lanham, MD: Rowman & Littlefield, 2005).

to regain its capabilities.[49] Vietnamese confidence in the Saigon government's ability to protect them nonetheless eroded.[50] Watching scenes of the fighting on their televisions, the American public became skeptical of Westmoreland and President Lyndon B. Johnson's assurances that the United States was winning in the Republic of Vietnam.[51] Fighting continued for several weeks in the former imperial capital at Hue and the Marine outpost at Khe Sanh.[52] At Khe Sanh and the surrounding hills, the enemy put to the test the strongpoint defense that USMACV implemented in mid-1967. Marines took a defensive stance, not patrolling more than 400 meters from their positions and relying on artillery and air support to engage the enemy troops bombarding the firebase.[53]

As the end of the 1968 monsoon season neared, Cushman proposed and Westmoreland accepted a three-phase counteroffensive in April with a combined Marine, U.S. Army, and Army of the Republic of Vietnam relief of Khe Sanh. An attack into the demilitarized zone and a raid into the A Shau Valley was to follow. Operation Pegasus, the relief of Khe Sanh that began on 1 April 1968 and lasted for two weeks, had an enduring impact on 3d Marine Division as the genesis of an innovative way to fight in I Corps. Operating from a new base at Ca Lu called Landing Zone Stud, 16 kilometers east of Khe Sanh, the 1st Cavalry Division (Airmobile) used airmobility to advance along Route 9 while 1st Marines cleared the road below. By blasting landing zones out of the jungle on hilltops, 1st Cavalry Division helicopters leapfrogged artillery and soldiers closer to Khe Sanh and screened the advancing 1st Marines. The linkup occurred on 8 April. The Marines declared Route 9 open three days later, and Operation Pegasus, the largest III MAF operation to that point in the war, concluded on 15 April.[54]

TOWARD HIGH MOBILITY

Major General Raymond Davis watched Operation Pegasus with interest. As deputy commanding general of Provisional Corps, Vietnam, the newly created corps-level headquarters subordinate to III MAF, Davis had a unique vantage point.[55] He believed Marine operations in Vietnam were too defensive minded in theory and plodding in execution.[56] He abhorred the logic behind strongpoint defenses, thinking them akin, as he put it, to "hiding out" and an abandonment of aggressive pursuit and destruction of the enemy, the cornerstones of Marine training.[57] By comparison, the U.S. Army units within the Provisional Corps—the 1st Cavalry Division (Airmobile) and 101st Airborne Division—operated almost exclusively with helicopters, allowing them extraordinary maneuverability. They eschewed dictates about what constituted a proper landing zone and dropped their troopers where most advantageous. This had the potential not only to open up the battlefield, as helicopters could quickly insert units en masse in territory that NVA units considered safe, but to wrest the initiative away from the enemy.

General Westmoreland created the Provisional Corps to integrate operations in I Corps and improve inter-Service communication via a Joint U.S. Army and Marine Corps staff.[58] To that end, the commanding general of the Provisional Corps, Army Lieutenant General William B. Rosson, selected Davis as his deputy. He also chose Davis for personal reasons, as the two men were close friends while students at the National War College in Washington, DC, 10 years prior.[59] Immediately upon Davis's arrival at Provisional Corps headquarters, Lieutenant General Rosson stressed to his dep-

---

[49] Ngo Vinh Long, "The Tet Offensive and Its Aftermath," in *The American War in Vietnam*, ed. Jayne Werner and David Hunt (Ithaca, NY: Southeast Asia Program, Cornell University, 1993), 23–45.
[50] Daddis, *Westmoreland's War*, 144.
[51] Chester J. Pach Jr., "The War on Television: TV News, the Johnson Administration, and Vietnam," in *A Companion to the Vietnam War*, ed. Marilyn B. Young and Robert Buzzanco (Malden, MA: Blackwell Publishers, 2002), 450–69.
[52] Col Richard D. Camp Jr., USMC (Ret), *Death in the Imperial City: U.S. Marines in the Battle for Hue, 31 January to 2 March 1968*, Marines in the Vietnam War Commemorative Series (Quantico, VA: Marine Corps History Division, 2018).
[53] Simmons, "Marine Corps Operations in Vietnam, 1968," 113.

[54] Col Richard D. Camp, USMC (Ret), with LtCol Leonard A. Blasiol, *Ringed by Fire: U.S. Marines and the Siege of Khe Sanh, 21 January to 9 July 1968*, Marines in the Vietnam War Commemorative Series (Quantico, VA: Marine Corps History Division, 2019), 77–84; and Shulimson et al., *The Defining Year, 1968*, 284.
[55] USMACV renamed Provisional Corps to XXIV Corps in August 1968.
[56] Shulimson et al., *The Defining Year, 1968*, 308.
[57] Raymond Davis, *The Story of Ray Davis: General of Marines* (Fuquay-Varina, NC: Research Triangle Publishing, 1995), 195.
[58] Cosmas, *MACV*, 75–76.
[59] Davis, *The Story of Ray Davis*, 188.

General Raymond G. Davis

Defense Department (Marine Corps) A7018919

Raymond Gilbert Davis was born 13 January 1915 in Fitzgerald, Georgia. He grew up in Atlanta, graduating from a technical high school in 1933, where he spent three years as a member of the Army Reserve Officers Training Corps (ROTC) unit. He remained in Army ROTC when he enrolled at Georgia Tech (Georgia Institute of Technology) to pursue a degree in chemical engineering. Upon graduation, Davis commissioned into the Army Reserves but instead accepted an appointment as a Marine officer. He reported to The Basic School at the Philadelphia Navy Yard in June 1938, becoming friends with Gregory "Pappy" Boyington, future Marine Corps fighter ace in World War II. Davis's company commander at The Basic School was then-captain Lewis "Chesty" Puller. It was the first of several encounters with Puller throughout Davis's career.[1]

Davis spent the first year out of The Basic School aboard the USS *Portland* (CA 33). After receiving weapons and artillery instruction at Quantico, Virginia, and Aberdeen, Maryland, Davis spent the interwar years in the 1st Antiaircraft Machine Gun Battery of the 1st Marine Division. In August 1942, Davis landed at Guadalcanal as commanding officer of the battery. By October 1943, he was in command of 1st Special Weapons Battalion, 1st Marine Division. Six months later, he became commanding officer of 1st Battalion, 1st Marines, 1st Marine Division. It was in this capacity that he received the Navy Cross for actions at Peleliu. In November 1944, he returned to the United States and was assigned to Marine Corps Schools at Quantico.[2]

When Davis embarked for Korea in August 1950, he was the commanding officer of 1st Battalion, 7th Marines. Between the two wars, he served with 1st Provisional Marine Brigade on Guam for almost two years. Davis received the Medal of Honor for his actions during the Battle of the Chosin Reservoir, where his battalion opened a mountain pass to allow two trapped regiments to escape. By the time he left Korea for an assignment at Headquarters Marine Corps, he had received a Medal of Honor, two Silver Stars, a Legion of Merit with combat "V," and a Bronze Star with combat "V."[3]

---

[1] Davis, *The Story of Ray Davis*, 30–31.
[2] Davis, *The Story of Ray Davis*, 75–90.
[3] James H. Willbanks, ed., *America's Heroes: Medal of Honor Recipients from the Civil War to Afghanistan* (Santa Barbara, CA: ABC-CLIO, 2001), 73.

After several assignments at Headquarters Marine Corps, with stints back at Quantico and a year at the National War College, Davis went to Headquarters, United States European Command, for three years as part of the intelligence staff. In 1963, he was promoted to brigadier general and assigned to 3d Marine Division as assistant division commander. A year later, he returned to Headquarters Marine Corps as assistant director of personnel and then assistant chief of staff, G-1. In March 1968, he went to the Republic of Vietnam to serve as deputy commanding general, Provisional Corps. Two months later, he became commanding general, 3d Marine Division, a position he held until April 1969.[4]

Davis returned to Quantico in May 1969 and served in a variety of billets until becoming commanding general, Marine Corps Development and Education Command upon his promotion to lieutenant general in July 1970. In February 1971, President Richard M. Nixon nominated Davis to Assistant Commandant of the Marine Corps. A year later, he retired from active duty.[5] He spent his retirement in Georgia and dedicated his time and energy to Korean War veteran issues. Davis died of a heart attack on 3 September 2003.[6]

uty the need to improve "the effectiveness of forces."[60] The two spent considerable time discussing the merits of airmobility. Rosson, like much of USMACV, lamented the defensive posture of 3d Marine Division, believing that it gave an advantage to the enemy.[61] For a month, Davis followed his boss around the area of operations with his own helicopter, observing first Operation Pegasus and then Operation Delaware.[62]

With the influence of Rosson, his reading of U.S. Army doctrine, and experience watching airmobility in action, Davis conceived of a Marine adaptation of the concept. He sent his aide, Captain Richard D. Camp, to visit the units of the Provisional Corps in preparation of an article the two coauthored and submitted to the *Marine Corps Gazette* on how Marine regiments could conduct heliborne assaults, with the intention of influencing the Fleet Marine Force.[63] When "Marines in Assault by Helicopter" appeared in the *Marine Corps Gazette* in September 1968, units were already practicing its core tenets in Vietnam, which Davis later summarized as: pick a pinnacle and knock off the top, put in reconnaissance to observe the landing zone, put in engineers to clear the vegetation, prepare artillery positions, have infantry patrol down the hills rather than up them, and repeat the process to create mutually supporting fire support bases.[64] Put even more simply: high mobility.

Marines practiced high mobility in fall 1968 because Davis was in a position to directly influence operations. On 22 May 1968, after only two months at Provisional Corps, he took command of 3d Marine Division. Two hours after the change-of-command ceremony, with a draft of "Marines in Assault by Helicopter" in his hand, he assembled his staff officers and regimental commanders and notified them that high mobility was now the guiding principle of the division.[65] The first opportunity to test the concept came at the end of the month, when the People's Army of Vietnam's *308th Division*, fresh replace-

---

[4] ALMARS Number 053/03, "Death of General Raymond G. Davis, Former Assistant Commandant of the Marine Corps," 3 September 2003.
[5] Davis, *The Story of Ray Davis*, 231–52.
[6] Richard Goldstein, "Gen. Raymond Davis, War Hero, Dies at 88," *New York Times*, 5 September 2003.

[60] Shulimson et al., *The Defining Year, 1968*, 308.
[61] Davis interview.
[62] Simmons, "Marine Corps Operations in Vietnam, 1968," 114.
[63] Col Richard D. Camp, USMC (Ret), "Taking Command: A Lesson in Leadership," *Marine Corps Gazette* 83, no. 6 (June 1999): 79.
[64] MajGen Raymond G. Davis and Capt Richard D. Camp, "Marines in Assault by Helicopter," *Marine Corps Gazette* 52, no. 9 (September 1968): 22–28. See also notes, Command and Staff College, "Notes from Previous Dewey Canyon Symposiums," 2 June 1993, box 56, folder 1, Raymond G. Davis Papers, HD Archive.
[65] Davis, *The Story of Ray Davis*, 194.

ments from Hanoi who relieved the mauled *304th Division*, appeared south of Khe Sanh, near a salient in the border. In Operations Robin North and Robin South, 3d and 4th Marines engaged the new arrivals. The operations were successful not just in killing and capturing 725 enemy troops and seizing large amounts of weapons and equipment but proving that high mobility worked. Two reinforced Marine regiments with eight batteries of artillery used mountaintop fire bases and helicopter resupply to bludgeon a fresh NVA division, forcing it to return to the Democratic Republic of Vietnam to refit.[66] Though the sample size was small, Davis's high mobility concept appeared applicable to Marine operations in the mountains of I Corps.

Davis afterward sought to apply the concept to his entire division. He instituted a series of operational changes that stressed unit cohesion, support, and maneuverability. Infantry and artillery battalions returned to their parent regiments, as did all support units. The division closed any combat base that did not serve an operational function, and no more than a reinforced company defended those that remained. To Davis's mind, two mobile battalions could cover the same area as five static battalions.[67] Regiments now freed up from containing enemy forces undertook deep offensive, preemptive operations. The reconnaissance effort in the field increased to 60 four-man teams, with 20 operating at any given time to ensure a steady flow of information.[68] As a new tactical mode, high mobility emphasized making infantry maneuverable through cutting ties to static defenses and lines of communication while keeping it protected—in other words, taking artillery wherever infantry went. Fire support bases were the foundation of the concept. Perched atop easily defendable hilltops and ridgelines, they provided a protective artillery fan to infantry units and mutually supporting fires to the batteries. The crucial element of fire support bases and thus high mobility, however, was sourcing and maintaining enough helicopter lift capability to build, occupy, and resupply the hilltop positions.[69]

Douglas A. Yeager Collection, Archives Branch, Marine Corps History Division

Illustrating the Joint nature of high mobility, a U.S. Army Boeing Vertol CH-47 Chinook heavy-lift helicopter carries a 105mm howitzer from Battery A, 1st Battalion, 12th Marines, during Operation Robin South, June 1968.

REFINEMENT OF STRATEGY IN I CORPS

The new concept was indeed an adaptation of the U.S. Army's airmobility, but it was not based on tenets absent from Marine doctrine. Beginning in the 1940s, the Marine Corps developed what it termed *vertical envelopment*, an alternative way to ensure assault troop mobility without massing ships off shore. For a variety of reasons, from technology to funding, the air assault concept did not fully mature but the assumptions remained.[70] In 1968, Marines like Major General Davis saw a correlation between vertical envelopment and airmobility.[71] It was also clear to some Marines that fire support bases were akin to inland beachheads and that they gave 3d Marine Division and Task Force Hotel the ability to project combat power

---

[66] Simmons, "Marine Corps Operations in Vietnam, 1968," 120.
[67] Command and Staff College, "Notes from Previous Dewey Canyon Symposiums."
[68] Simmons, "Marine Corps Operations in Vietnam, 1968," 121; and Davis, *The Story of Ray Davis*, 197–98.
[69] Maj Robert V. Nicoli, "Fire Support Base Development," *Marine Corps Gazette* 53, no. 9 (September 1969): 38–43.

[70] LtCol Eugene W. Rawlins, USMC, *Marines and Helicopters, 1946–1962*, ed. Maj William J. Sambito, USMC (Washington, DC: History and Museums Division, Headquarters, U.S. Marine Corps, 1976).
[71] Davis interview.

Gen Creighton W. Abrams (right) talks with MajGen George I. Forsythe, commanding general of the U.S. Army's 1st Cavalry Division (Airmobile).

when assaulting islands of enemy strength.[72] The Joint concept for amphibious operations reasoned that the usefulness of such operations stemmed "from mobility and flexibility," or "the ability to concentrate balanced forces and to strike with great strength a selected point in the hostile defense system."[73] In western Quang Tri Province, Davis's primary difficulty was traversing rugged, unfamiliar terrain to reach enemy sanctuaries, all while maintaining lines of communication. Treating the assault as an amphibious operation allowed him to bypass a series of hazards and select the most advantageous locations to attack.

Built into amphibious operation doctrine, too, was the intended effect of compelling the enemy to disperse their forces, resulting in "making expensive and wasteful efforts in attempting to defend [their] coast line."[74] If 3d Marine Division could elicit such behavior from the enemy, it would force North Vietnamese troops to stand and fight in western Quang Tri Province, not the populated lowlands. High mobility offered such a high degree of maneuverability, though, that 3d Marine Division did not have to focus solely on destroying enemy units. It could instead target the enemy's ability to wage war in the Republic of Vietnam by attacking base areas and cutting off units already in the field, allowing friendly forces elsewhere to destroy them. It was, in effect, a realization of the strategy that Westmoreland tried to implement in 1967. The difference, however, was that the offensive-minded high mobility attacked rather than interdicted the enemy.

In June 1968, Army General Creighton W. Abrams Jr. replaced Westmoreland. As Westmoreland's recent deputy, Abrams previously had operational control over III MAF after the creation of USMACV Forward at Phu Bai in February 1968. Like his former boss, Abrams believed American and Republic of Vietnam forces should stop NVA units at the border before they could reach the population centers. Both, too, skirmished with Marine leadership over how to fight in I Corps, which spilled into the press in the United States. Where both men diverged, however, was the strongpoint defense.[75]

On 19 June, 3d Marine Division initiated Operation Charlie, the evacuation and eventual razing of Khe Sanh. The 1968 spring offensive indicated an American willingness to conduct operations with mobility in mind, but the bulldozing of Khe Sanh, the last vestige of the strongpoint defense system that Major General Davis termed a "$6 billion blunder," announced it demonstrably.[76] Hanoi claimed a political and psychological victory, as the Americans left a base that commanders in Saigon once declared they would defend at all costs. USMACV and III MAF, however, argued that defen-

---

[72] Col Marion C. Dalby, "Task Force Hotel's Inland Beachheads," *Marine Corps Gazette* 53, no. 1 (January 1969): 34–38.
[73] *Doctrine for Amphibious Operations, 1 August 1967*, LFM-01 (Washington, DC: Department of the Army and the Navy, 1967), 1-3.
[74] *Doctrine for Amphibious Operations*, 1-3. See also LtGen Herman Nickerson Jr., *Leadership Lessons and Remembrances from Vietnam* (Washington, DC: History and Museums Division, Headquarters, U.S. Marine Corps, 1988), 78.
[75] Gregory A. Daddis, *Withdrawal: Reassessing America's Final Years in Vietnam* (New York: Oxford University Press, 2017), 17–19.
[76] Davis, *The Story of Ray Davis*, 8.

Defense Department (Marine Corps) 14-1677-68

A Marine from 3d Battalion, 4th Marines, looks back at what remains of the Khe Sanh airstrip before leaving.

sive outposts of the McNamara Line, with Khe Sanh as the linchpin, were no longer necessary. Changes on the battlefield necessitated adjustments to force structure and posture: more troops, helicopters, and firepower available in I Corps allowed III MAF to confront the increased enemy strength in the tactical zone. For Abrams, it was a "whole 'nother ball game now." Rather than battalions tied to outposts like Khe Sanh that were doing nothing but "listen[ing] to AFN [American Forces Network] to get the daily reports on all the fighting going on other places," they used mobility to concentrate on the enemy.[77] This pleased Lieutenant General Cushman, who advocated for more mobile operations since becoming III MAF commander in June 1967. Now, his forces could range Quang Tri Province, operating from Landing Zone Stud, the 1st Cavalry Division base during Operation Pegasus. The Marines turned the landing zone into a combat base, named it after the 18th Commandant of the Marine Corps, General Alexander A. Vandegrift, and prepared to attack into enemy sanctuaries.[78]

Operations from Vandegrift Combat Base would have to wait, though, as the enemy weighted their operations toward Da Nang in August 1968, using Viet Cong guerrillas to breach defenses that NVA units could then exploit. Battalions from 1st and 5th Marine Divisions, along with troopers from the U.S. Army's 1st Air Cavalry Division, blunted those attempts,

---

[77] Transcript, J-2/J-3 Briefing for General Creighton Abrams and Lieutenant General Ferdinand J. Chesarek, 22 January 1969, in *Vietnam Chronicles: The Abrams Tapes, 1968–1972*, ed. Lewis Sorley (Lubbock: Texas Tech University Press, 2004), 111, hereafter J-2/J-3 Briefing for Abrams and Chesarek.

[78] Simmons, "Marine Corps Operations in Vietnam, 1968," 121.

Company D, 1st Battalion, 7th Marines, moves across the Vu Gia River at the start of Operation Mameluke Thrust, 24 May 1968.

Defense Department (Marine Corps) A371543

however. In the late 1968 Operations Allen Brook and Mameluke Thrust, American forces handed Hanoi another tactical failure, pushing the enemy to their sanctuaries in the mountains of northwestern I Corps, along the Laotian border.[79]

In fall 1968, 3d Marine Division pursued NVA units across I Corps to areas west and south of Khe Sanh, where no Marine or soldier operated before. On 28 November, 9th Marines conducted Operation Dawson River, the first in a series of operations that swept the mountains in and near enemy Base Area 101.[80] Lasting until Christmas 1968, Operation Dawson River confirmed what 3d Marine Division believed—that the Da Krong Valley was a primary infiltration route of the People's Army of Vietnam's *7th Front*, the headquarters that controlled all enemy troops within Quang Tri Province.[81] As a result, 9th Marines expanded operations between Khe Sanh and the Laotian border on 1 January 1969, in Operation Dawson River West. Contact with the enemy was light in both operations, mostly NVA regiments screening a larger force. Enemy presence was obvious, however, as the Marines uncovered numerous caches, some of which had American equipment lost during or after Khe Sanh. It was clear to 9th Marines' leadership that they were keeping the enemy on the move, which, while less than ideal, still yielded the tactical

---

[79] Simmons, "Marine Corps Operations in Vietnam, 1968," 122.
[80] Shulimson et al., *The Defining Year, 1968*, 455.
[81] 9th Marines Command Chronology (ComdC), 1 December to 31 December 1968, HD Archive.

Marines from 3d Platoon, Company A, 1st Battalion, 9th Marines, move through elephant grass during Operation Dawson River West, 11 December 1968.

benefit of knocking the opposing force off-balance.[82] After 12 months of costly offensives, NVA and Viet Cong leaders preferred to avoid direct confrontation and relied instead on small unit harassment closer to the population centers. Eventually, Major General Davis hoped to use high mobility to descend on substantial enemy base areas and disrupt NVA operations not just on the border but throughout Quang Tri Province.

Davis's performance in spring and summer 1968 pleased Abrams, who considered the division commander a "brilliant, professional tactician" as well as "one of these very quiet, self-effacing, modest men."[83] Personal characteristics aside, Davis earned Abrams's praise because high mobility transformed the tactical situation in Quang Tri Province. The increased mobility centered at Vandegrift Combat Base allowed III MAF and USMACV to free up units for operations in I Corps and elsewhere in the Republic of Vietnam. Only the year before, Abrams opined, if "you'd mentioned the 9th Marines being [in western Quang Tri Province], they'd have run you out of the country as being crazy, get a psychiatrist after you."[84] With a new concept in place, III MAF was expanding its control of Quang Tri Province with each operation.

At the end of 1968, the Marines, who began the year sitting in their defensive positions waiting to interdict enemy troops, were now conducting large-scale attacks into North Vietnamese base areas. There were plenty of targets from which to choose. At the beginning of 1969, III MAF estimated there were 90,000 enemy troops in I Corps or in the

---

[82] Barrow interview.
[83] J-2/J-3 Briefing for Abrams and Chesarek, 111–12.

[84] J-2/J-3 Briefing for Abrams and Chesarek, 111.

3D MARINE DIVISION OUTPOSTS
JANUARY 1969

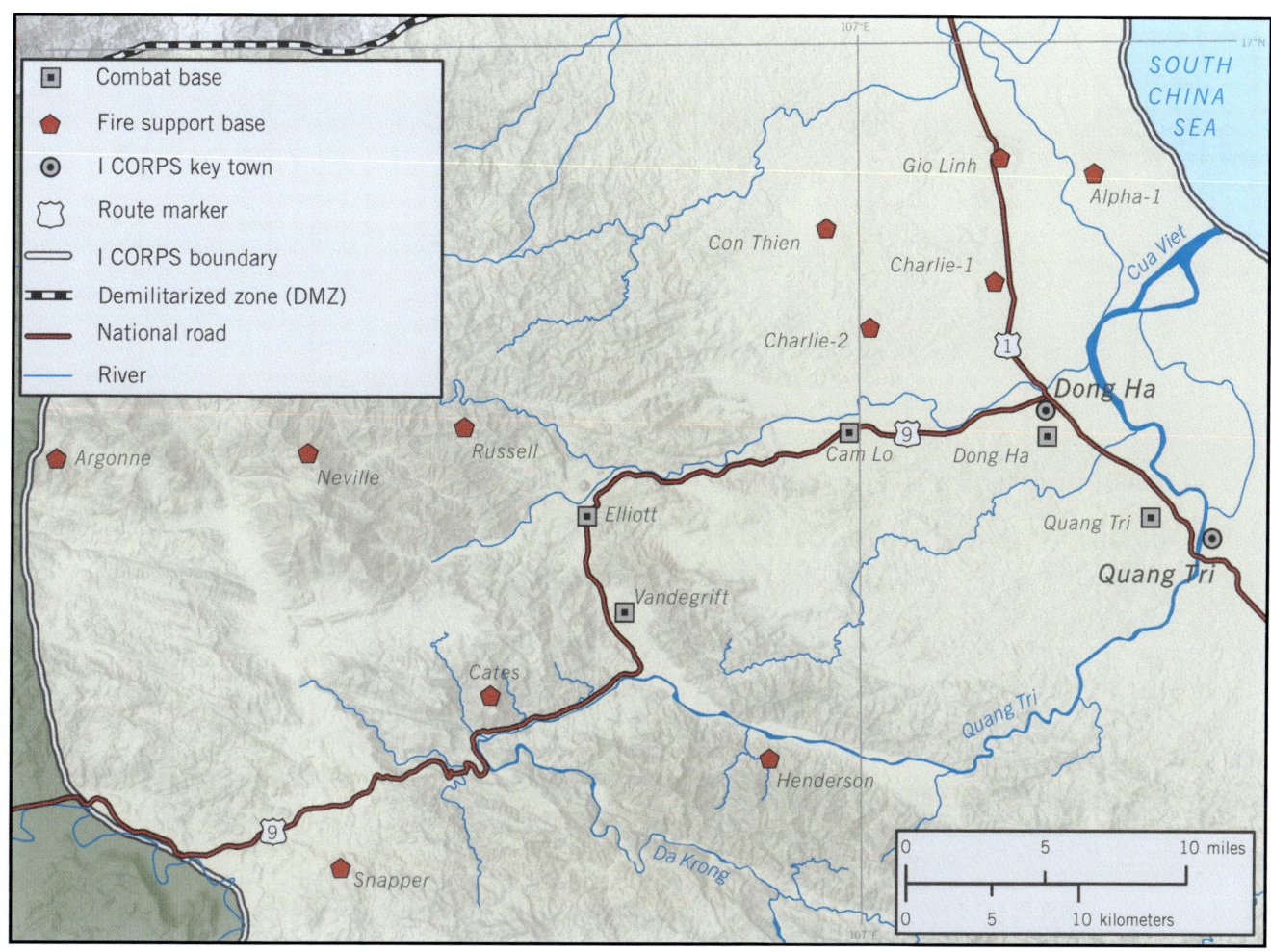

Map courtesy of Pete McPhail, adapted by MCUP

demilitarized zone and 89 battalions inside the tactical zone.[85] The task now was to ensure that they could go no farther.

PLANNING AND PHASE I–II OPERATIONS
PLANNING

Planning for the upcoming operation into the upper Da Krong Valley, called Operation Dawson River South, fell to 9th Marines and its commanding officer, Colonel Robert H. Barrow, in conjunction with the commanding officer of 2d Battalion, 12th Marines, Lieutenant Colonel Joseph R. Scoppa Jr. Their task was to block the infiltration route between Laos and the A Shau Valley, denying the enemy an administrative and supply base from which to conduct a new Tet Offensive.[86] Planning only took five days, an achievement owing to Colonel Barrow's capable staff, who spent the previous five months honing their knowledge of mountain warfare in Quang Tri Province.[87]

---

[85] BGen Edwin H. Simmons, "Marine Corps Operations in Vietnam, 1969–1972," in *The Marines in Vietnam, 1954–1973*, 132.

[86] Intelligence Annex, Task Force Hotel to 9th Marines, "Operation Order 2-69 (Dawson River South)," 14 January 1969, box 9, folder 4, Vietnam War Collection, HD Archive.

[87] Barrow interview.

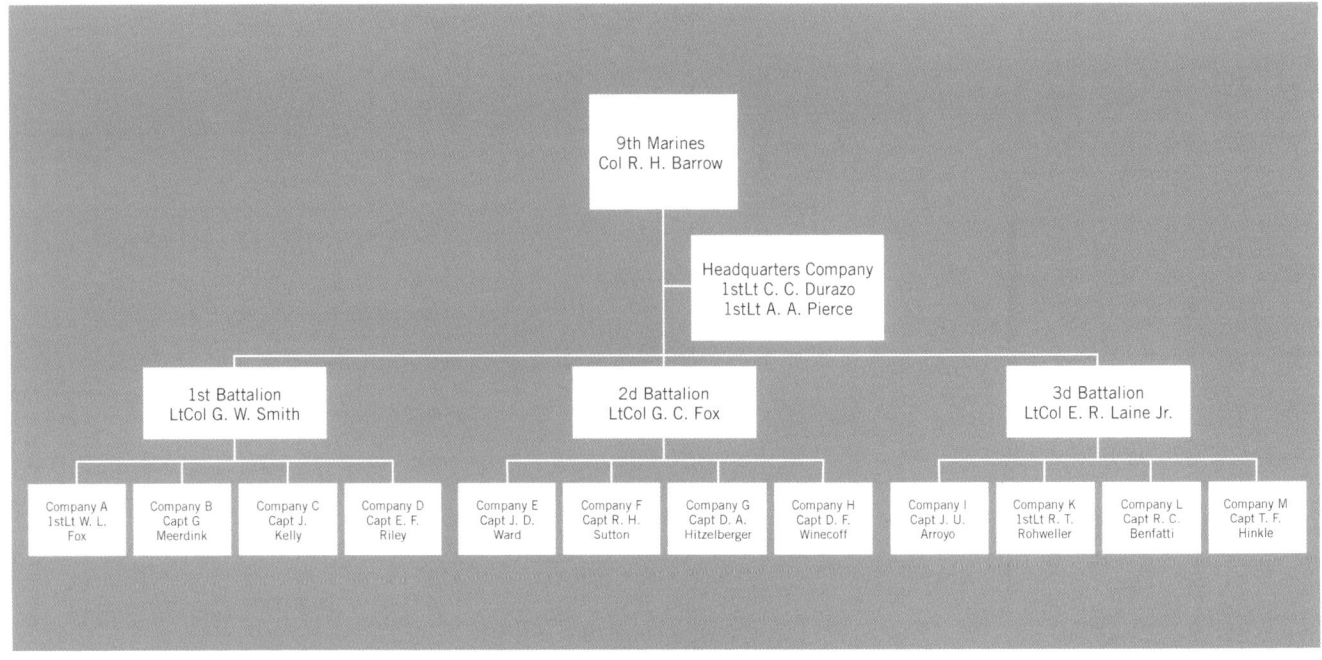

Courtesy of author, adapted by MCUP

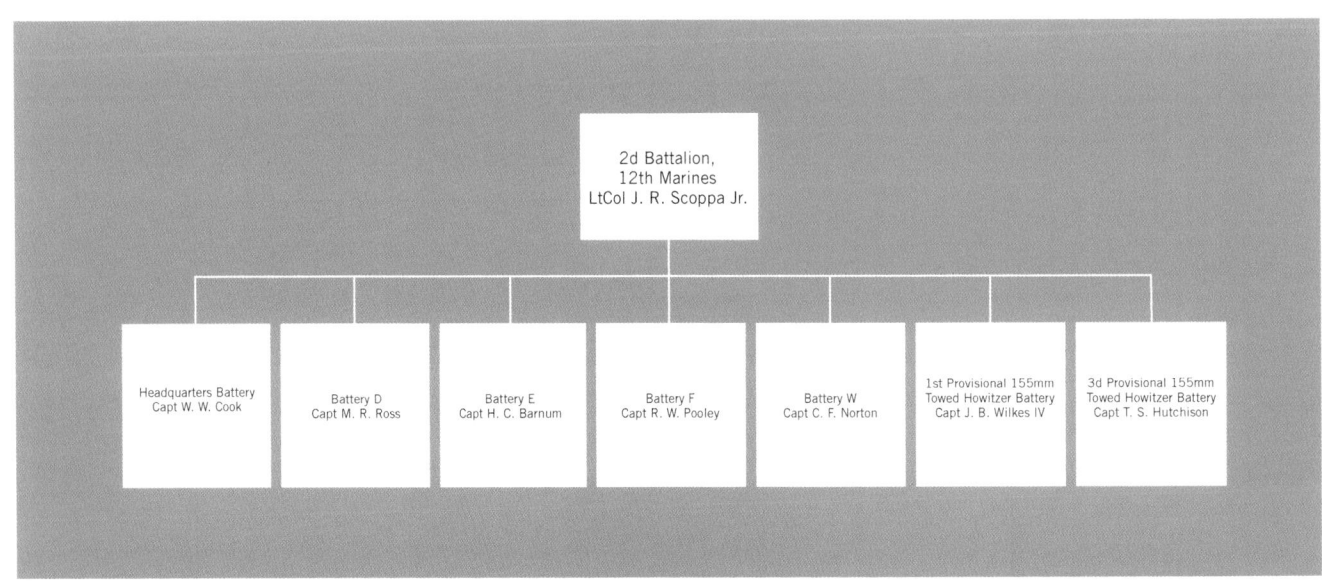

Courtesy of author, adapted by MCUP

Aerial reconnaissance photograph of the Da Krong River.
Archives Branch, Marine Corps History Division

Barrow and his battalion commanders knew from reconnaissance flights that location, terrain, and weather would make operating in the area difficult, much like the A Shau Valley. To the north was the Da Krong Valley's high, jungle-covered mountains and steep ridgelines. The battalions would have to fight everywhere from 200 meters above sea level near the Da Krong River up to 1,200 meters in the hills and mountains and contend with everything from elephant grass taller than the average person to triple-canopied jungle. The southern extent of the area of operations was the rugged Laotian border. Where the Marines would target Base Area 611, the border ran west-east, preventing any flanking maneuver and leaving a regimental attack with battalions on line as the only option. The enemy controlled most of the critical high ground along the border. On the western end of the area of operations was the north-south running Co Ka Leuye (Hill 1175). Whoever occupied it could bring down fires on much of the border area. On the eastern end was Tam Boi (Hill 1224) and Tiger Mountain (Hill 1228). Of particular importance was Tiger Mountain, which dominated not just the Da Krong Valley to its northwest but also the northern approaches to the A Shau Valley to its east. The operation's proximity to the Laotian border concerned Barrow and his commanders. They knew

NVA units could use Laos as a sanctuary and employ artillery against targets inside the Republic of Vietnam. It would be difficult for 9th Marines and 2d Battalion, 12th Marines, to determine the size of opposing forces, let alone close with and destroy the enemy.[88]

Unpredictable weather was a complicating factor as well, since the operation took place during the final months of the northwest winter monsoon. A persistent drizzle and average temperatures between 60 and 70 degrees Fahrenheit during the day would chill Marines, and heavy ground fog and low-hanging clouds would shroud the mountains and hills in the mornings.[89] The conditions would make it difficult to depend on fixed-wing support as well as helicopter resupply and casualty evacuation to Vandegrift Combat Base, the vital lifeline upward of 40 kilometers away.[90]

Apart from the challenges of location, terrain, and weather, Colonel Barrow and his staff factored in a lack of intelligence and time. As near as III MAF could tell, the People's Army of Vietnam's *6th Regiment*, *9th Regiment*, *65th Artillery Regiment*, and *83d Engineer Regiment* moved through the Da Krong and A Shau Valleys to two base areas: Base Area 101, southwest of Quang Tri, and Base Area 114, west of Hue.[91] They presumed from the existence of antiaircraft artillery in the area that NVA forces protected something of value. There was no time to find out just what that was beforehand and little time to conduct the operation itself. Major General Davis ordered Colonel Barrow to begin the operation "as soon as practicable" after 22 January but finish it by 12–14 February in anticipation of renewed enemy attacks during Tet. When the holiday came, Davis wanted the greatest number of battalions on hand as possible to respond to NVA and Viet Cong attacks throughout Quang Tri Province.[92]

---

[88] Barrow to Davis report.
[89] LtCol George W. Smith to Col Robert H. Barrow, "Combat Operations After Action Report," 31 March 1969, box 41, file 2432056, Entry UD 07D 1, Command Chronologies and Related Documentation, Records of the U.S. Marine Corps, RG 127, National Archives and Records Administration (NARA), hereafter Smith to Barrow report.
[90] Barrow interview.
[91] Charles R. Smith, *U.S. Marines in Vietnam: High Mobility and Standdown, 1969* (Washington, DC: History and Museums Division, Headquarters, U.S. Marine Corps, 1988), 27.
[92] Commanding Officer's Planning Guidance, "Dawson River South—Upper Dakrong Valley," 9th Marines, no date, box 41, file 2432056, Entry UD 07D 1, RG 127, NARA.

DA KRONG VALLEY

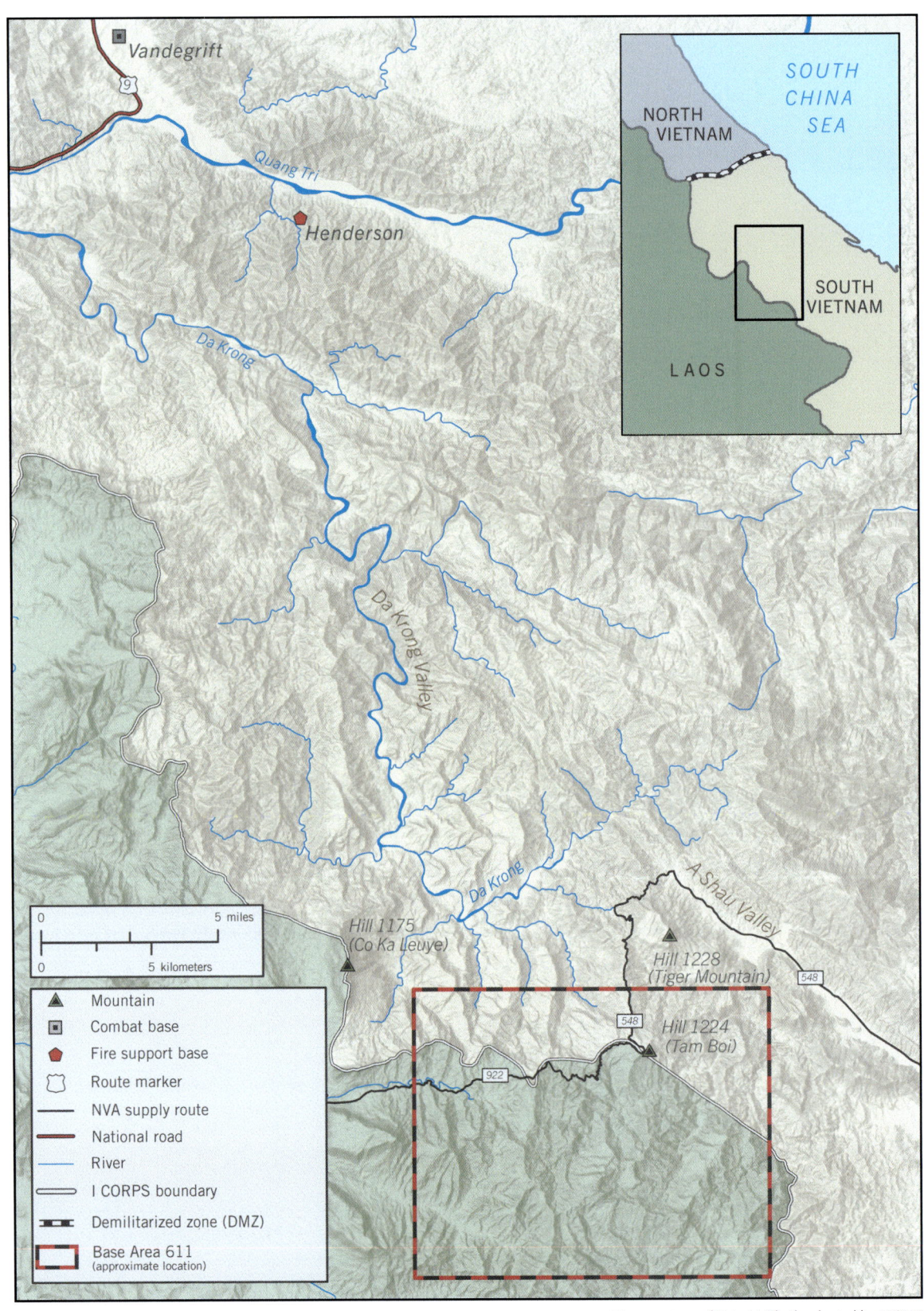

Map courtesy of Pete McPhail, adapted by MCUP

Douglas A. Yeager Collection, Archives Branch, Marine Corps History Division

A Boeing Vertol CH-46 Sea Knight medium-lift helicopter from Marine Medium Helicopter Squadron 364 (HMM-364), the "Purple Foxes," offloads Marines.

After five days of discussions with 12th Marines, 1st Marine Aircraft Wing (1st MAW), and XXIV Corps headquarters, Barrow and his staff presented a plan built around high mobility.[93] It called for a three-week, three-phase operation with three infantry battalions progressing south from Vandegrift Combat Base through the Da Krong Valley. Rifle companies were to venture out from fire support bases and operate independently, when necessary. Artillery batteries on hilltops would provide an eight-kilometer overlapping fan of protection for the maneuver units. As the Marines moved south and outran their fire support, they would set up new fire support bases and repeat the process. In Phase I, Barrow wanted to establish his regiment's presence in the area. His Marines would either reoccupy or build six new outposts on a north-south line—the first of which was 12 kilometers southeast of Vandegrift Combat Base—and stock them with enough artillery ammunition to support an infantry battalion for 10 days.[94] In a supporting operation beginning on 29 January to the east of the A Shau Valley called Operation Sherman Peak, a battalion from 2d Brigade, 101st Airborne Division, and a battalion from 3d Regiment, 1st ARVN Division, would clear the area south of Base Area 114. Phase II was a shift to the attack. The 9th Marines would conduct sweeps around their positions and clear the enemy from the bases that would become the regiment's rear when pushing farther south. Phase III departed somewhat from the high mobility concept, as it was a more conventional regiment-in-the-attack operation. Its objective was Tiger Mountain and Tam Boi, the high ground that split the Da Krong and A Shau Valleys. Taking those would deny the enemy the ability to transit troops and supplies from Laos into their valley sanctuary. Barrow and his planners contemplated a heliborne assault near 9th Marines' ultimate objectives, but he reasoned that the threat of enemy antiaircraft defenses made an overland attack more attractive.[95] The thrust into the Da Krong Valley, then, would be a bold operation. Not only were the Marines heading into inhospitable terrain during the monsoon season, but they were also stepping into an enemy refuge with prepared defenses.

PHASE I

The first phase that became Operation Dewey Canyon began as Operation Dawson River South, which was a reorientation for units still in the field conducting Operation Dawson River West. To secure the approaches for later phases, Boeing Vertol CH-46 Sea Knight medium-lift helicopters, with escorts of North American Rockwell OV-10 Bronco light attack and observation aircraft and Bell UH-1E Iroquois gunships, lifted elements of 9th Marines' 1st and 3d Battalions and 2d Battalion, 12th Marines, to three separate fire support bases from Vandegrift Combat Base and its environs. The assault troop lift took three days to complete, with squadrons such as Marine Medium Helicopter Squadron 262 (HMM-262) racking up 54 flight hours during 162 sorties to move all of 3d Battalion, 9th Marines' 676 combat troops.[96] The helicopter squadrons inserted the units equidistant from each other, on top of the ridgelines that ran southeasterly toward the A Shau Valley.

---

[93] Barrow to Davis report.
[94] Fire Support Annex, Task Force Hotel to 9th Marines, "Operation Order 2-69 (Dawson River South)," 14 January 1969, box 9, folder 4, Vietnam War Collection, HD Archive.

[95] Barrow to Davis report.
[96] Marine Medium Helicopter Squadron 262 ComdC, 1 January to 31 January 1969, HD Archive.

General Robert H. Barrow

Defense Department (Marine Corps) A707991

Robert H. Barrow was born on his family's historic Rosale Plantation north of Baton Rouge, Louisiana, on 5 February 1922. Upon graduating high school in 1939, he enrolled at Louisiana State University in Baton Rouge. The Japanese attack on Pearl Harbor occurred his junior year, and though he had aspirations to join the U.S. Army, he signed up for the Marine Corps' Platoon Leaders Course after seeing a Marine recruiter on campus. Impatient to get to war, Barrow requested active duty in November 1942. He left his degree unfinished and went to recruit training in San Diego, California. By March 1943, he was in officer training at Quantico, Virginia.

Barrow spent all of World War II in China, where he led a group of Chinese guerrillas in sabotaging Japanese infrastructure in Hunan Province as part of the Sino-American Cooperative Organization. After the war, he first joined the Shore Patrol in Shanghai and then was aide to Major General Keller E. Rockey, III Amphibious Corps commander for two and a half years. In Korea as commander of Company A, 1st Battalion, in Colonel Lewis B. "Chesty" Puller's 1st Marines, Barrow first saw action at Inchon on 15 September 1950. A week later, he and his Marines defended an important crossroads at Yongdung-po against an enemy mechanized unit, action for which he received the Silver Star. During the Chosin River campaign, Company A defended the critical Funchilin Pass when it took the high ground above the road, ensuring the 1st Marine Division could escape to Hungnam. For his actions there, Barrow received the Navy Cross.

A short stint at Headquarters Marine Corps awaited Barrow after returning home, but he was soon back in Asia, this time for covert operations on small islands in the Yellow Sea. For the next 10 years, he held a series of billets ranging from Marine officer instructor at Tulane University in New Orleans to the Publications Branch of the Landing Force Development Center at Quantico. Barrow spent much of the 1960s in the Pacific. His time as the G-3 plans officer in Lieutenant General Victor H. Krulak's Fleet Marine Force, Pacific, earned him a Legion of Merit. He made more than 30 trips to Vietnam in three years before taking command of 9th Marines. He commanded 9th Marines for nine months and finished his tour as deputy G-3 of the III Marine Amphibious Force. His time in Vietnam yielded a Bronze Star, Legion of Merit, and the Army's Distinguished Service Cross for Operation Dewey Canyon.

Following the war, Barrow received a promotion to brigadier general and served as commanding general of Marine Corps Bases at Okinawa, where he earned a third Legion of Merit. His experience as commanding general of Marine Corps Recruit Depot, Parris Island, in South Carolina led him to raise recruiting standards when he became deputy chief of staff for manpower at Headquarters Marine Corps. In October 1976, he became commanding general of Fleet Marine Force, Atlantic. Two years later, Barrow was Assistant Commandant, a post he held until 29 June 1979, when he became the 27th Commandant of the Marine Corps. He served as Commandant until his retirement on 30 June 1983. Barrow died on 30 October 2008.[1]

---

[1] BGen Edwin H. Simmons, "Robert Hilliard Barrow, 1979–1983," *Commandants of the Marine Corps*, ed. Allan R. Millett and Jack Shulimson (Annapolis, MD: Naval Institute Press, 2004), 437–56.

Marines arrived at Fire Support Base Henderson first, a position along a gently sloping, long, and narrow hilltop, 12 kilometers southeast of Vandegrift Combat Base and at the mouth of the Da Krong Valley. At 1000 on 18 January, the command post for Lieutenant Colonel Elliott R. Laine's 3d Battalion, 9th Marines, along with the 81mm mortar platoon, 106mm recoilless rifle platoon, and Companies I and K, boarded helicopters after sweeping the Ba Long Valley on the regiment's left flank. They joined Battery F, 2d Battalion, 12th Marines, which arrived at Fire Support Base Henderson on 14 January.[97] By nightfall, the battalion command post, a section from the mortar platoon, the recoilless rifle platoon, and Company K were in position.[98] On 20 January, Company L, 3d Battalion, 9th Marines, reoccupied Fire Support Base Tun Tavern, which lay atop a hill on an opposing ridgeline from Fire Support Base Henderson, eight kilometers to the southwest. The batteries of 2d Battalion, 12th Marines, last occupied the triangular-shaped fire support base the month prior and left it in good condition on Lieutenant Colonel Scoppa's assumption that they may return to any firebase in the area of operations.[99]

Soon, guns from Battery D, 2d Battalion, 12th Marines, fired at the next objective seven kilometers southeast, called Fire Support Base Shiloh, which lay on the same range as Fire Support Base Tun Tavern.[100] At 0600 on 21 January, CH-46s helilifted Company A from Lieutenant Colonel George W. Smith's 1st Battalion, 9th Marines, to Fire Support Base Shiloh. The "Walking Dead," as the Marines in the veteran unit referred to themselves, were to secure the hill and investigate suspected enemy activity in the area.[101] After First Lieutenant Wesley L. Fox's Company A took the fire support base, Battery E, 2d Battalion, 12th Marines, arrived to provide the next eight-kilometer fan of artillery fire. When night fell on 21 January, the Marines prepared for the next push into the

---

[97] 2d Battalion, 12th Marines ComdC, 1 January to 31 January 1969, HD Archive.
[98] LtCol Elliott R. Laine Jr. to Col Robert H. Barrow, "Combat After Action Report," 27 January 1969, attached to 3d Battalion, 9th Marines ComdC, 1 January to 31 January 1969, HD Archive.
[99] LtCol Joseph R Scoppa Jr. interview (no interviewer or date), Marine Corps Oral History Collection, hereafter Scoppa interview.
[100] Smith, *High Mobility and Standdown, 1969*, 30.
[101] LtCol George W. Smith to Companies B, C, D, and H&S, fragmentary order, attached to 1st Battalion, 9th Marines ComdC, 1 January to 31 January 1969, HD Archive.

9th Marines

The story of 9th Marines has multiple endings and beginnings. Commandant George Barnett ordered the formation of the regiment on 10 November 1917 as part of the expansion of the Marine Corps during World War I. The regiment's first deployment was a month later to Guantánamo Bay, Cuba, where they assisted 7th Marines in counterinsurgency operations. After seven months, the regiment sailed to Galveston, Texas, to protect oil shipments against German subterfuge, where it remained until the end of the war. On 25 April 1919, it disbanded in Philadelphia, Pennsylvania.[1]

To ensure preparedness for any future conflict, the Marine Corps reorganized 9th Marines on 1 December 1925 as a reserve regiment, with its component parts spread throughout the Midwest. The Marine Corps disbanded the regiment once again on 1 September 1937. During the expansion for World War II, the Marines reactivated the regiment on 12 February 1942. Originally part of the 2d Marine Division, 9th Marines transferred to 3d Marine Division and trained in California until sailing for New Zealand on 24 January 1943.[2] The young regiment's first battle was at Bougainville, where it proved itself in combat. Seven months later, 9th Marines assaulted Guam. After refitting and training, the regiment arrived at Iwo Jima as part of the floating reserve. When it arrived on the island, it became the 3d Marine Division's spearhead, taking Airfield No. 2, breaking the Japanese main line of resistance in the Motoyama Plateau, and securing the northeastern shore. The war ended when the regiment was refitting on Guam.[3]

On the last day of 1945, the Marine Corps disbanded 9th Marines at Camp Pendleton, California. In autumn 1947, the Marine Corps began reorganizing regiments with impressive combat traditions and reputations, but at battalion strength. The 2d Battalion, 5th Marines, became 9th Marines on 1 October 1947 and sailed for Guam, where it remained for a year until the evacuation of American nationals and dependents from China after the collapse of the Chinese Nationalist Party in the country's civil war. When 9th Marines returned to the United States in May 1948, it became part of 2d Provisional Marine Regiment at Camp Lejeune, North Carolina. Due to further reorganization, the Marine Corps redesignated the regiment 3d Battalion, 6th Marines, and dropped 9th Marines from the muster rolls on 17 October 1949.[4]

By the Korean War, the familiar pattern for 9th Marines repeated. On 17 March 1952, the Marine Corps reactivated the regiment. It did not see combat but traveled to Japan to serve as a mobile force in readiness in August 1953 before being stationed permanently in Okinawa. For the next eight years, 9th Marines participated in nearly every major exercise in Asia. The battalion landing team from 3d Battalion, 9th Marines, came ashore at Red Beach on 8 March 1965 to protect the airfield at Da Nang, Republic of Vietnam. By August 1965, the entire regiment was in Vietnam, where it remained for the next four years, participating in operations across much of eastern and northern I Corps. It was one of the first regiments to rotate out of Vietnam

---

[1] Truman R. Strobridge, *A Brief History of the 9th Marines* (Washington, DC: Historical Branch, G-3 Division, Headquarters, U.S. Marine Corps, 1967), 1–2.
[2] Strobridge, *A Brief History of the 9th Marines*, 2–3.

[3] Strobridge, *A Brief History of the 9th Marines*, 3–13.
[4] Strobridge, *A Brief History of the 9th Marines*, 13–14.

during the withdrawal, returning to Okinawa in July 1969. The regiment's 2d Battalion was the reaction force in the Mayaguez Incident in May 1975. The names of three of the battalion's Marines lost during the Mayaguez Incident are some of the last enshrined on the Vietnam Veterans Memorial in Washington, DC.[5]

The regiment's battalions spent much of the 1970s and 1980s involved in training and humanitarian operations in the Pacific. The 3d Battalion fought in the 1990–91 Gulf War, serving as part of Task Force Papa Bear in taking the Kuwait International Airport in February 1991.[6] In late 1992 and early 1993, 2d and 3d Battalions deployed to Somalia for Operation Restore Hope.[7] In September 1994, the Marine Corps once again disbanded 9th Marines. The regiment remained absent from the muster rolls until the Global War on Terrorism. The Marine Corps activated 9th Marines' battalions independent from one another between April 2007 and May 2008. All three deployed to Iraq in support of Operation Iraqi Freedom (2003–11) and Afghanistan in support of Operation Enduring Freedom (2001–14). As part of the Service's drawdown, the Marine Corps deactivated one battalion per year, from 2013 to 2015. The 9th Marines' colors remain cased today.

---

[5] Strobridge, *A Brief History of the 9th Marines*, 14–20.
[6] Paul W. Westermeyer, *U.S. Marines in the Gulf War, 1990–1991: Liberating Kuwait* (Quantico, VA: Marine Corps History Division, 2014), 170–71.
[7] Col Dennis P. Mroczkowski, USMCR (Ret), *Restoring Hope: In Somalia with the Unified Task Force, 1992–1993* (Washington, DC: History Division, Headquarters, U.S. Marine Corps, 2005).

Da Krong Valley. Fighting was limited so far. In clearing operations around Fire Support Bases Henderson, Tun Tavern, and Shiloh, Marines saw more evidence of enemy activity than they did actual North Vietnamese troops, though most was old. The closer they inched toward the Laotian border, however, the more likely they would run into the main body of enemy forces they knew existed.

With 1st and 3d Battalions, 9th Marines, securing the approaches to the area of operations, and batteries from 2d Battalion, 12th Marines, perched atop key terrain to provide fire support, Lieutenant Colonel George C. Fox's 2d Battalion, 9th Marines, could enter the fight. They arrived at Vandegrift Combat Base from the field on 19 January—after Operation Dawson River West ended that day—and spent the next two days preparing to spearhead the next phase of the operation. At 0950 on 22 January, 2d Battalion, 9th Marines, boarded CH-46s at Vandegrift Combat Base and set out for two missions. Companies E and H would go eight kilometers southeast of Fire Support Base Shiloh and assault a 600-meter hilltop, soon to be dubbed Fire Support Base Razor, a name which Colonel Barrow selected in honor of Major General Davis's nickname. Companies F and G would go five kilometers farther to take a new landing zone, Landing Zone Dallas, which guarded the Marines' western flank.[102]

PHASE II

These moves initiated the second phase in the operation to block enemy transit in and out of the A Shau Valley. During planning, 9th Marines determined that their ultimate objective, and indeed the main thrust of their mission, was the area considerably south of where 3d Marine Division outlined in their 14 January message to Task Force Hotel.[103] Given that they intended to expand the area of operations beyond what Task Force Hotel envisioned in the original planning guidance, 9th Marines initiated a change in codename for the action taking place on 22 January. The push into the Da Krong Valley now would be known as Operation Dewey Canyon.[104]

---

[102] Smith, *High Mobility and Standdown, 1969*, 31; and Air Annex, Air Schedule for Operation Dewey Canyon, attached to 9th Marines ComdC, 1 January to 31 January 1969, HD Archive.
[103] Barrow interview.
[104] Units remaining at Fire Support Bases Henderson and Tun Tavern, however, still operated under the codename Dawson River South. Smith, *High Mobility and Standdown, 1969*, 33.

TACTICAL SITUATION, PHASE I

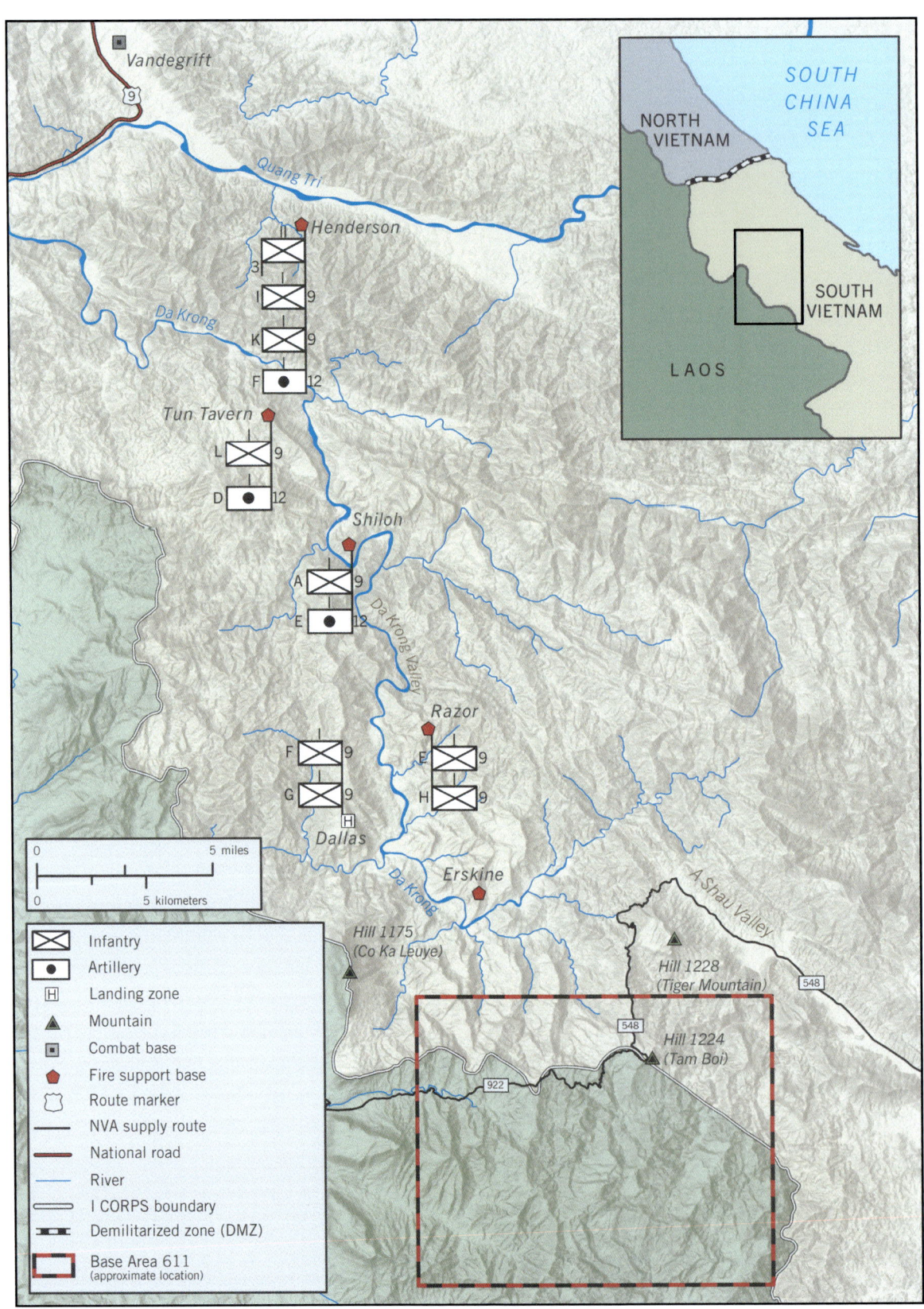

Map courtesy of Pete McPhail, adapted by MCUP

Marines from 2d Battalion, 9th Marines, wait at Vandegrift Combat Base for helicopters to take them to what would become Fire Support Base Razor and the high ground surrounding it.

Fire Support Base Razor and Landing Zone Dallas were the first positions that Marines constructed in the operation. Opening Fire Support Base Razor on time was a particular feat of ingenuity and cooperation. Shortly after Company H, 2d Battalion, 9th Marines, set foot on the hilltop that fixed-wing aircraft bombed to blast room for helicopters to land, its commanding officer, Captain David F. Winecoff, realized that few of his Marines knew how to swing an axe. Almost immediately, the Marines splintered 50–60 percent of the axe handles, a distinct problem for the company designated to fell trees on a hilltop that was to become home of the regimental command group. Engineers from Company C, 3d Engineering Battalion, stepped in to assist felling trees up to one meter in diameter. Using a combination of chainsaws and the remaining axes, the infantrymen cleared enough of Fire Support Base Razor for the engineers to begin construction.[105]

CH-46s next flew in a bulldozer and other equipment from detachments of Support Company, 3d Engineer Battalion, and Company C, 11th Engineer Battalion, to expand the landing zone and construct gun pits as well as ammunition berms.[106] Unbeknownst to the engineers, however, someone prematurely declared the support base operational. Before the bulldozer

---

[105] Capt David F. Winecoff, interview with SSgt Willis S. Bernard Jr., 5–9 March 1969, Marine Corps Oral History Collection, hereafter Winecoff interview.
[106] Barrow to Davis report.

Defense Department (Marine Corps) 3D-2-3056-69

Marine engineers chop down trees to clear a landing zone.

Defense Department (Marine Corps) A371975

Bulldozers like this Case M450, which were light enough that helicopters could lift them, proved vital during Operation Dewey Canyon. While engineers felled trees, operators cleared areas and dug gun pits so fire support bases could open quickly.

operator could finish the landing zone, helicopters descended with external loads for the regimental command post. Equipment, gear, and supplies clogged the tiny ridgeline and blocked the bulldozer. After several hours of effort, the Marines cleared the landing zone and helicopters brought Battery F, 2d Battalion, 12th Marines, whose commander added to the engineers' frustrations. Artillerymen attempted to occupy positions the engineers were frantically in the process of building, setting off tense exchanges between officers.[107] Adding to the confusion was a near-constant convoy of helicopters, setting down 46 tons of cargo and 1,544 Marines. Despite the bedlam, the battery made their 105mm howitzers operational by nightfall on 23 January.[108] The next day, the regimental command group boarded helicopters at Vandegrift Combat Base and set off for Fire Support Base Razor, as did Lieutenant Colonel Scoppa's Headquarters Battery.[109] When night came on 24 January, 9th Marines were in place in the zone of action.[110]

On the morning of 25 January, the regiment repeated its scheme of maneuver from the previous few days, this time with components of the battalion that initiated the entire operation. At 0935, helicopters lifted three companies from 3d Battalion, 9th Marines, to an area six kilometers southeast of Fire Support Base Razor, on the same series of hills that overlooked the Da Krong River. Once again, air strikes pummeled the area to level trees and push back any enemy presence. The three landing zones the Marines secured clung to a

---

[107] Company C, 3d Engineer Battalion to 3d Engineer Battalion, "After Action Report: Operation Dewey Canyon (22 January–3 March 1969)," 18 March 1969, HD Archive, hereafter Company C to 3d Engineer Battalion Report.

[108] LtGen Keith B. McCutcheon, "Marine Aviation in Vietnam, 1962–1970," in *The Marines in Vietnam, 1954–1973*, 287.

[109] Scoppa interview.

[110] Smith, *High Mobility and Standdown, 1969*, 31; and Barrow to Davis report.

A Sikorsky CH-53 Sea Stallion heavy-lift helicopter lands on the narrow ridgeline of Fire Support Base Razor, home of Battery F, 2d Battalion, 12th Marines, on 22 January.

jagged ridgeline named Co Ka Va that stretched more than 1,100 meters. Within short order, an advance party of artillerymen and engineers arrived and began turning part of Co Ka Va into another fire support base, this one named Fire Support Base Cunningham after the first Marine aviator, Lieutenant Colonel Alfred A. Cunningham. By the end of 29 January, five of 2d Battalion, 12th Marines' six firing batteries were on Fire Support Base Cunningham. Joining Batteries D and E was Battery W, with their 107mm "Howtar" heavy mortars, and the 1st and 3d Provisional Batteries' towed 155mm howitzers.[111] With the batteries in place, Barrow had an artillery position in the center of his planned area of operations—one large enough that he could place an entire battalion above the valley. The fan of artillery coverage now extended to 11 kilometers for the 105mm and nearly 15 kilometers for the 155mm howitzers, which meant 9th Marines would have fire support to the edge of where it operated.[112]

Vital to the success of Operation Dewey Canyon thus far was Marine Aviation. CH-46s and Sikorsky CH-53 Sea Stallion heavy-lift helicopters gave 9th Marines and 2d Battalion, 12th Marines, the ability to move across more than 25 kilometers of rugged terrain, much of it unfamiliar territory, and reopen or construct 5 vital fire support bases. Most of this was through the efforts of a "Zippo" team, made up of representatives from 1st Marine Aircraft Wing and 3d Marine Division, with infantry, engineer, and helicopter and observation

---

[111] Scoppa interview. The M-98 was a 107mm (4.2-inch) heavy mortar mounted on a 75mm pack-howitzer chassis, giving it the name Howtar. It was light enough to be towed behind vehicles or airlifted into landing zones. The Marine Corps developed the concept with heliborne operations in mind: Inserting Whiskey Batteries, as they were called in Vietnam, with an infantry battalion provided the ground combat element commander immediate direct support.

[112] 12th Marines ComdC, 1 January to 31 January 1969, HD Archive. See also Smith, *High Mobility and Standdown, 1969*, 31.

Defense Department (Marine Corps) A192655

A Bell UH-1E utility helicopter from Marine Light Helicopter Squadron 367 (HML-367) lands at Fire Support Base Cunningham, 26 January 1969. In the foreground are Marines from Battery D, 2d Battalion, 12th Marines, and their 105mm howitzers.

aircraft specialists represented.[113] Answerable to the overall ground commander, the team coordinated helicopter assaults, provided landing zone and fire support, and assisted in base selection and preparation.[114] Joint assistance from U.S. Army Aviation also played a crucial role in the early days of the operation. Given that the weather was favorable for flying, Boeing Vertol CH-47 Chinooks and Sikorsky CH-54 Tarhes aug-

---

[113] The term *zippo* refers to a zone interpretation, planning, preparation, and overfly team. See Smith, *High Mobility and Standdown, 1969*, 87.

[114] McCutcheon, "Marine Aviation in Vietnam, 1962–1970," 287.

mented Marine Aviation, particularly when establishing Fire Support Bases Razor and Cunningham.

With the regiment now in the area of operations, 9th Marines cleared around the two southernmost fire support bases. On 24 and 25 January, companies from 2d and 3d Battalions swept the new territory, not only to destroy any enemy but to secure the regiment's flanks and maneuver themselves into position for Phase III. That new position, called Phase Line Red, was along the banks of the Da Krong River, where the stream flows east-west. On 9th Marines' right flank was 2d Battalion. On the left was 3d Battalion, operating from Fire Support Base Cunningham. After 2d and 3d Battalions were in place, 1st Battalion could move south from Fire Support Base Shiloh and take position in the middle of Phase Line Red for what Barrow envisioned as a regimental attack overland.[115]

Individual objectives for companies were integral to support the forthcoming phase. Company F, 2d Battalion, 9th Marines, would construct a new fire support base against Phase Line Red named Fire Support Base Erskine for General Graves B. Erskine, commanding general of 3d Marine Division during the Battle of Iwo Jima. The new position would be four kilometers southwest of Fire Support Base Cunningham and seven kilometers from the Laotian border. Company G, 2d Battalion, 9th Marines, was to seize Co Ka Leuye, a flat and broad ridgeline that straddled the border and towered over the flood plain below. This dominating position allowed the Marines to see beyond Fire Support Base Cunningham to the northeast, where Company K, 3d Battalion, 9th Marines, would take Landing Zones Lightning and Tornado. The two landing zones represented the eastern flank of the operation and from where ARVN soldiers would block any enemy escape into the A Shau Valley. Once the battalions maneuvered into their attack positions, 9th Marines would launch Phase III.[116]

After a week of operations, the regiment was far enough south that units ran into small groups of enemy soldiers while conducting sweeps around the fire support bases. On 25 January, while 3d Battalion, 9th Marines, and Battery D, 2d Battalion, 12th Marines, hacked away at what became Fire Support Base Cunningham, Company E, 2d Battalion, 9th Marines, made contact with an NVA force in the late morning, resulting in one enemy killed. Four hours later, a reconnaissance team codenamed Desert Sands skirmished with another group of soldiers that resulted in two more enemy killed. Much of the same played out the next several days, with Marines steadily engaging larger concentrations of troops, most of which appeared to be screening elements for the larger force across the border and along Route 922. Among them were guides, porters, and soldiers who tended to a communication system that, though primitive compared to American standards, was a credit to the resourcefulness and skill of the NVA. Company M, 3d Battalion, 9th Marines, discovered a four-strand telephone line that began in Laos and went all the way to Base Area 101, located in the Hai Lang Forest Preserve, nearly 30 kilometers east. The telephone line was invisible from the air, as enemy troops maintained the tree's foliage and strung the wire with glass insulators in the lower limbs.[117] Rather than destroy the communications wire, a five-man team tapped it, broke the enemy code, and intercepted traffic between the two base areas for several days.[118]

The battalions also encountered enemy defensive positions as they cleared the hills overlooking the Da Krong River. On the morning of 28 January, Captain Joseph U. Arroyo sent two platoons from his Company I, 3d Battalion, 9th Marines, to conduct patrols around Hill 561. The lead element from the northern patrol discovered a bunker complex with eight soldiers cooking breakfast. The Marines' attempt to capture the enemy soldiers turned into a firefight with an entire NVA platoon. Once the enemy force received reinforcements after two hours of contact and attempted to flank the Marines, the 3d Platoon commander radioed Captain Arroyo and requested reinforcements of his own. When Arroyo arrived at the bunker complex with two squads, he ordered artillery and air strikes. For the next hour, Marines from Company I retrieved their one dead and seven wounded while fixed-wing aircraft dropped ordnance on the enemy positions. Arroyo then ordered an assault on the bunkers, forcing the NVA troops

---

[115] 1stLt Gordon M. Davis, "Dewey Canyon: All Weather Classic," *Marine Corps Gazette* 53, no. 7 (July 1969): 36.
[116] Davis, "Dewey Canyon," 36.

[117] Davis, "Dewey Canyon," 36.
[118] LtCol Elliott R. Laine Jr., interview with SSgt Marshall Neal Jr., 13 April 1969, Marine Corps Oral History Collection, hereafter Laine interview.

TACTICAL SITUATION, PHASE II

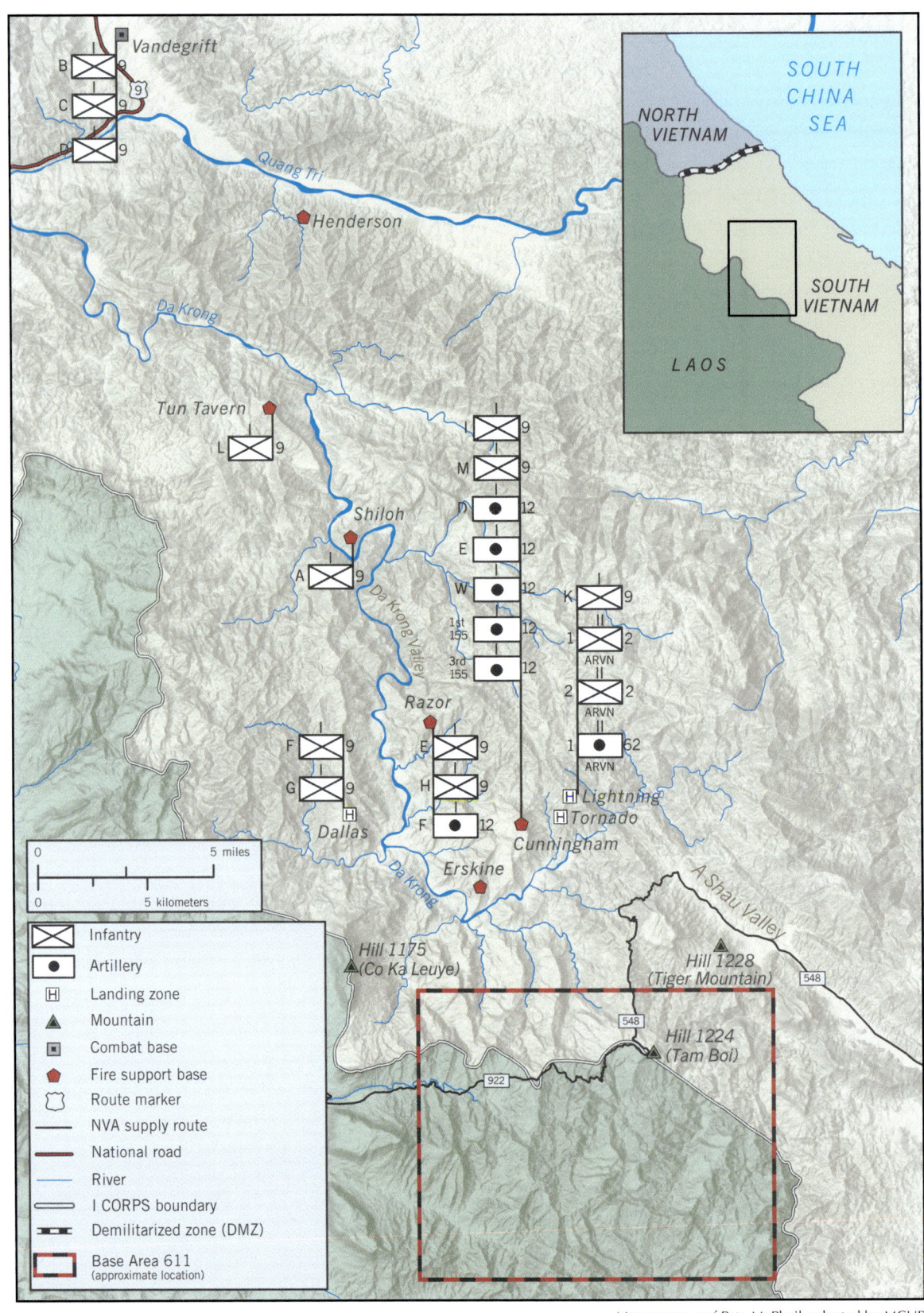

Map courtesy of Pete McPhail, adapted by MCUP

Captain Joseph U. Arroyo
Silver Star Citation

The President of the United States of America takes pleasure in presenting the Silver Star to Captain Joseph U. Arroyo (MCSN: 0-90455), United States Marine Corps, for conspicuous gallantry and intrepidity in action while serving as Commanding Officer of Company I, Third Battalion, Ninth Marines, THIRD Marine Division in connection with combat operations against the enemy in the Republic of Vietnam. On 28 January 1969, while participating in Operation DEWEY CANYON southwest of Quang Tri City, the Third Platoon of Company I sustained several casualties when it became pinned down by a large North Vietnamese Army force occupying well-concealed emplacements and employing hand grenades, small arms and automatic weapons. Fearlessly leading his men to the embattled area, Captain Arroyo observed that because of the thick vegetation it was impossible to utilize the full firepower of the combined units and that the enemy had fallen back to well-prepared positions from which they were utilizing crew-served weapons, thereby achieving fire superiority. Unwilling to risk additional casualties, Captain Arroyo directed the Marines to re-deploy to more tenable positions and, while providing rear security for the movement, spotted a seriously wounded man lying in proximity to the hostile emplacements. With complete disregard for his own safety, he rushed to the side of the casualty and removing his own protective jacket, placed it over his companion to prevent further injury from the enemy rounds impacting around him. While thus engaged, he sustained a serious wound in his arm from fragments of a hostile grenade but refusing to leave the wounded man, enlisted the aid of a nearby Marine to drag the casualty to a covered location. Steadfastly refusing medical attention, he then resolutely moved to a dangerously exposed location to adjust supporting artillery fire on the enemy entrenchments until he was certain that the now reinforced North Vietnamese unit could not mount an attack on the maneuvering Marine elements. After receiving first aid, Captain Arroyo maneuvered his men and the newly arrived First Platoon into attack positions then, despite the pain of his wounds led an aggressive assault against the enemy. Inspired by his fearless leadership and bold courage, the Marines overran the North Vietnamese Army emplacements, killed twenty-one hostile soldiers, and captured several weapons and a quantity of ammunition and documents containing information of intelligence value. By his courage, heroic actions, and unfaltering devotion to duty in the face of grave personal danger, Captain Arroyo defeated a determined enemy force which was attempting to infiltrate that section of the A Shau Valley under his cognizance and upheld the highest traditions of the Marine Corps and of the United States Naval Service.

to break contact and retreat south. For his actions, Arroyo received the Silver Star.[119]

With the discovery of units came proof of how well the North Vietnamese troops established the southern Da Krong Valley as a base of operations. Companies from 2d and 3d Battalion, 9th Marines, found a number of structures, from classrooms and barracks to bunkers, as well as a considerable number of fighting holes and booby traps. One of the first significant discoveries was the People's Army of Vietnam's *88th Field Hospital*, which Company F, 2d Battalion, 9th Marines, found near the Da Krong River when sweeping the area around Fire Support Base Razor and Landing Zone Dallas. An entire complex, the hospital had eight permanent buildings, including an operating room, and enough space for more than 150 patients. It was stocked with Soviet-made stainless-steel surgical instruments as well as 500 pounds of medical supplies.[120] Marines elsewhere learned how extensive the enemy's food source was. Near the hospital, 2d Battalion, 9th Marines, found a well-cultivated, six-acre corn field and mod-

---

[119] Capt Joseph U. Arroyo, interview with SSgt Willis S. Bernard Jr., 19 March 1969, Marine Corps Oral History Collection.

[120] Barrow to Davis report.

ern farming implements. Cleverly, NVA support troops staggered their planting schedule, ensuring portions of the field would be ready for harvest every few weeks to provide fresh food to the screening elements and the steady flow of soldiers transiting between Laos and the A Shau Valley. Apart from this large field along the Da Krong River, Marines also found gardens of squash and other vegetables and small plots of tobacco. To deny the enemy the food source, the Marines arranged for ARVN troops to harvest the crops and supply them to nearby civilians as part of a civic action program.[121]

INCLEMENT WEATHER SETS IN

The enemy's elusiveness from 19–30 January suggested that any pitched battles would come nearer Routes 922 and 548, and ultimately Base Area 611. The 9th Marines began a race for time when maneuvering into place for the climax of the operation. In his instructions, Colonel Barrow advised the battalions that the January and February weather in the Da Krong Valley might play a factor in the regiment's ability to operate. Forecasts predicted drizzle and fog, with only a slight chance for isolated thunderstorms. If anything, the planners believed, the extent of the weather would be a temporary inconvenience that would affect resupply and air cover, not a potential worry that could necessitate a reevaluation of the phased plan.[122]

Weather conditions deteriorated on 1 February, as Captain Daniel A. Hitzelberger and his Company G, 2d Battalion, 9th Marines, climbed Co Ka Leuye. They set off from Landing Zone Dallas the evening before to begin their movement up the heights. En route, the company discovered a group of NVA soldiers lying in wait, hoping to draw the Marines into an ambush. Rather than engage the enemy, Captain Hitzelberger bypassed the ambush site and crossed a tributary of the Da Krong River near the foot of the mountain. By nightfall, the Marines advanced as far as 500 meters up the ridge. Morning brought significant difficulties. As the company ground their way up slopes averaging 65–75 degrees, they were forced at times to climb rock cliffs with ropes. When the drizzling rain came, their progress slowed even more, but they pushed

Jonathan F. Abel Collection, Archives Branch, Marine Corps History Division

Representative of a rifleman's life during Operation Dewey Canyon, Marines from 2d Battalion, 9th Marines, climb one of the many ridgelines in the Da Krong Valley.

up the mountain now slick with red mud. Intermittent heavy rain and shrouding fog reduced visibility to only 25 meters. It meanwhile became clearer to Barrow down at the regimental command post that Hitzelberger and his Marines might reach the summit only to be all alone, cut off with little hope of resupply or assistance.[123]

While Company G trudged up Co Ka Leuye, the other companies of the regiment attempted to secure the objec-

---

[121] 2d Battalion, 9th Marines ComdC, 1 January to 31 January 1969, HD Archive. See also Barrow to Davis report, 12–14.
[122] Intelligence Annex, Task Force Hotel to 9th Marines, "Operation Order 2-69 (Dawson River South)," 14 January 1969, box 9, folder 4, Vietnam War Collection, HD Archive.

[123] Smith, *High Mobility and Standdown, 1969*, 34.

tives vital to the impending Phase III attack. More than 10 kilometers away from Hitzelberger, Company F, 2d Battalion, 9th Marines, took what would become Fire Support Base Erskine on 1 February. Work began immediately, but the storm moved in and halted construction. The same series of events played out four kilometers east of Fire Support Base Cunningham, where Company K, 3d Battalion, 9th Marines, seized a mountaintop and began building Landing Zone Lightning. When the fog and drizzle arrived, the ARVN's 1st and 2d Battalion, 2d Regiment, and 1st Battalion, 62d Artillery Regiment, were on the ground. All helicopter operations halted. The 62d Artillery Regiment received just one 105mm howitzer and 400 rounds of ammunition.[124]

The enemy, who the Marines caught only fleeting glimpses of for more than a week, used the weather to announce their presence. Fire Support Base Cunningham was the first target. On 2 February, batteries from the People's Army of Vietnam's *65th Artillery Regiment* fired 30–40 rounds from Soviet-built D-74 122mm field guns. Marine artillerymen believed the enemy guns were inside Laos, out of range of American howitzers but close enough to exploit the D-74's 24-kilometer reach. In the barrage, a close round temporarily disabled one of the 155mm howitzers of 1st Provisional Battery, and a direct hit destroyed the fire direction center for 3d Provisional Battery.[125] The Marines of 2d Battalion, 12th Marines, rallied and 1st Provisional Battery's fire direction center quickly took the duties of the stricken unit, while Batteries D and E maintained counterbattery fire throughout the shelling despite the incoming rounds. When the barrage was over, there were five dead and five wounded Marines.[126] Lieutenant Colonel Scoppa later moved the 155mm howitzers to Fire Support Base Erskine to provide counterbattery capability in the southwestern portion of the area of operations.[127]

The situation at the fire support bases, though not dire, became less than ideal. The enemy's 122mm field guns harassed the hilltop positions sporadically, and it became a game of cat and mouse between NVA artillerymen and Marine aerial observers, with the former firing off rounds before the latter arrived on station and attempted—always in vain—to spot where the field guns were. Since the enemy ceased their firing whenever an observer was aloft, the Marines used loitering aircraft as an effective form of counterbattery fire.[128] Conserving rounds was necessary for 2d Battalion, 12th Marines, regardless. The batteries at the southernmost fire support bases received only their initial stock of ammunition from Marine CH-53 Sea Stallions and U.S. Army CH-47 Chinooks before the cloud banks and drizzle arrived on 1 February. When the bad weather did not lift for the next two days, Scoppa made a concerted effort to preserve what he had.[129]

Procedures that Major General Davis and his staff designed to make high mobility operations economical effectively doubled as weather contingencies for the Marines during Operation Dewey Canyon. In April 1968, 1st MAW formed Provisional Marine Aircraft Group 39 at Quang Tri Airfield to coordinate air operations with 3d Marine Division units without involving the wing at Da Nang. From Vandegrift Combat Base, Assistant Wing Commander Brigadier General Homer S. Hill had the authority to coordinate with his 3d Marine Division counterpart, Brigadier General Frank Garretson, who also served as commanding general of Task Force Hotel. This real-time cooperation mitigated issues that complicated support from HMM-262, Marine Medium Helicopter Squadron 161 (HMM-161), and Marine Observation Squadron 6 (VMO-6) and improved air-ground teamwork.[130]

Davis and his staff also attempted to diversify the logistics chain. At the beginning of Operation Dewey Canyon, the primary logistics support area was Vandegrift Combat Base, with Quang Tri Combat Base as an alternate resupply point. The poor weather over both the Da Krong Valley and the logistics support areas necessitated Task Force Hotel to request resupply missions from the 101st Airborne Division, located near Hue at Camp Evans, southeast of the Marine combat bases. Multiplying the number of logistics support areas for Operation Dewey Canyon, as the logic went, increased the opportunities for helicopters to exploit breaks in the weather over

---

[124] Smith, *High Mobility and Standdown, 1969*, 34; and Davis, "Dewey Canyon," 36.
[125] Provisional batteries were temporary units for special assignments.
[126] 2d Battalion, 12th Marines ComdC, 1 February to 28 February 1969, HD Archive.
[127] Scoppa interview.

[128] Barrow interview.
[129] Scoppa interview. See also Smith, *High Mobility and Standdown, 1969*, 34–35.
[130] Davis interview; and Davis, *The Story of Ray Davis*, 200.

A CH-53 approaches Fire Support Base Razor while a chaplain conducts a service. Note the representative low cloud cover in the Da Krong Valley below.

the resupply points as well as the Da Krong Valley. The location of the logistics support areas was crucial. Camp Evans gave helicopters an east-west axis into the area of operations, while Vandegrift and Quang Tri Combat Bases were on a north-south axis.[131] In the interim, 9th Marines and 2d Battalion, 12th Marines, relied on Lockheed C-130 Hercules transport aircraft and the TPQ-10 radar system for resupply drops, though sparingly, as the recoverable rate was only 40–50 percent.[132] In the area of operations, Fire Support Base Cunningham served as the central logistics hub for 9th Marines, with a regimental tactical-level logistic control group that coordinated with the battalions, operations officer, and air liaison officer. A miniature logistics support area at Fire Support Base Cunningham was capable of resupplying eight rifle companies per day.[133] Previously, Fire Support Base Shiloh served as the miniature logistics support area for the operation when it was the southernmost fire support base. There, Scoppa stored more than 7,000 rounds of ammunition in excess of normal stockage objectives.[134] Despite the preparations, difficulties still existed due to distance—the Da Krong Valley was between 40–50 kilometers from each logistics support area—as well as coordination and communication between the Marine air-ground

[131] Scoppa interview.
[132] Barrow to Davis report.
[133] Barrow to Davis report.
[134] Scoppa interview.

Photo courtesy of BGen Frank E. Garretson, USMC (Ret)

From left: BGen Frank E. Garretson, commanding general, Task Force Hotel; MajGen Raymond G. Davis, commanding general, 3d Marine Division; and BGen Homer S. Hill, assistant wing commander, 1st Marine Aircraft Wing.

team and U.S. Army helicopter companies supporting operations elsewhere in I Corps.[135] The overriding problem, however, was the unrelenting low-hanging clouds and drizzling rain.

At the 9th Marines command post, Barrow realized that he was at a decision point. From recent experience, he and his staff anticipated the weather improving within two or three days, and they could recommence the phased plan at the first chance. Conditions had not improved after four days, though. The companies that were to take the important objectives to support the regimental assault were caught before completing all of their assignments, and Barrow's battalions were not

abreast: Lieutenant Colonel Smith's 1st Battalion, 9th Marines, never reached its position on Phase Line Red between 2d and 3d Battalions, 9th Marines. Barrow and his staff reevaluated the situation and determined the weather would not improve during the next few days. He ordered the companies back to the fire support bases, where they could offer each other mutual protection and draw from stocked rations and a limited water supply. From Lieutenant Colonel Laine's 3d Battalion, 9th Marines, Companies I, M, and K joined Company L at Fire Support Base Cunningham. Lieutenant Colonel Fox's 2d Battalion, 9th Marines, was more spread out: Company H was on Fire Support Base Razor, Company F was on Fire Support Base Erskine, and Company E was at Landing Zone Dallas. Of all the units in the regiment, Company G was in

---

[135] Davis interview; Davis, *The Story of Ray Davis*, 200; and Barrow to Davis report.

At Vandegrift Combat Base, a CH-53 picks up a load of supplies from a stockpile.

the most precarious position, alone at the top of Co Ka Leuye and out of rations and water.[136]

FIGHTING ON CO KA LEUYE

After receiving the battalion order to return to safety, Captain Hitzelberger led his soggy, hungry Marines down the mountain on the morning of 5 February. Most of his men had not eaten in three days—and some four or five. The mist soaked them to the core and exhaustion showed on faces that now wore two-week-old beards.[137] With low-hanging clouds enshrouding them and saturated ground making every step precarious, Company G struggled down the mountain. They covered only 1,000 meters by 1300, climbing down the same cliff faces that they went up five days prior, using a variety of ropes to keep from slipping on the mud, rocks, and heavy undergrowth.[138] An hour later, they stopped altogether when the point element, 2d Squad from 3d Platoon, saw three enemy soldiers off to their right. They reported back to Hitzelberger, who decided to search the area. Hitzelberger knew there were hostile units on the mountain. During the past few nights, the enemy probed the company's perimeter, which accounted for

---

[136] Davis, "Dewey Canyon," 36.
[137] PFC Wade R. Cupps, statement, in LCpl Thomas P. Noonan Jr. Medal of Honor Recommendation, Biographical Files, Reference Branch Collection, HD Archive, hereafter Noonan recommendation.

[138] Capt Daniel A. Hitzelberger to Secretary of the Navy, award recommendation, 24 April 1969, in Noonan recommendation.

Jonathan F. Abel Collection, Archives Branch, Marine Corps History Division

A CH-46 from Marine Medium Helicopter Squadron 161 (HMM-161), the "Greyhawks," delivers supplies and ammunition on a sling. Note the mangled tree stump in the foreground, indicative of how Marines blasted fire support bases out of the jungle.

his Marines' lack of sleep. He ordered the column to remain in place and sent forward the point fireteam to discover if there was a larger force and, if possible, grab the three soldiers.

At the front of the fireteam was Private First Class Kenneth E. Jackson. While inching forward, Private First Class Jackson noticed a concealed enemy soldier and immediately notified his squad leader, who ordered the team to press ahead. When they moved forward, the Marines took sporadic sniper fire before small arms fire erupted in front of them. The team had walked into a U-shaped ambush, with enemy troops in bunkers and spider holes hidden among the terrain. The initial burst hit Jackson. He crumpled to the ground with wounds to both legs but returned fire. Exposed and bleeding, he poured suppressive fire into the enemy positions, allowing his teammates to seek cover.[139]

NVA soldiers had Jackson pinned down, and he was unable to move without being silhouetted against the foggy sky. The cloud cover limited visibility, and the Marines struggled to see anything beyond 10 meters. To aid his wounded rifleman, the 3d Platoon commander ordered forward Hospital Corpsman Third Class James D. Head and Lance Corporal E. F. Brower. Lance Corporal Brower went first, moving down the slope past Jackson, who was alive but severely wounded, to establish a firing position. He could see outlines of enemy soldiers in the blackness below and threw all four of his grenades at the shapes. His M16A1 rifle jammed after he fired several magazines. Jackson offered up his rifle, allowing Brower to continue laying down suppressive fire. Moments later, Brower heard Corpsman Head yell that Jackson was dead. A rifle grenade then exploded nearby and riddled the corpsman's leg with shrapnel. Brower crawled to Corpsman Head and worked on his wounds when another grenade exploded. Now both men had shrapnel wounds to the legs. Brower shouted for help. The company's senior medic, Hospital Corpsman Second Class Bruce B. Bernstein, ran down the hill but was hit with rifle fire in the chest and neck immediately after reaching Brower.

When Brower and Head advanced to tend to Jackson, Hitzelberger ordered 3d Platoon forward. He also told Second Lieutenant Jack B. Henderson to take his 2d Platoon and swing left. Henderson's men moved out but quickly hit the bulk of the enemy force. A hail of automatic weapon, small arms, and rifle grenade fire raked his platoon. The NVA soldiers were disciplined, firing at shadows and voices rather than spraying the jungle in front of them. They also aimed at legs, knowing that corpsmen and Marines would come to the aid of their wounded comrades and present more targets. The pinned-down platoons sheltered behind rocks to escape the withering fire and the snipers throwing grenades from trees.[140]

---

[139] Capt Daniel A. Hitzelberger, interview with SSgt Willis S. Bernard Jr., 5–9 March 1969, Marine Corps Oral History Collection, hereafter Hitzelberger interview.
[140] 2dLt Jack Henderson, statement, in Noonan recommendation; and Sgt Robert D. Gaudioso, interview with SSgt Willis S. Bernard Jr., 5–9 March 1969, Marine Corps Oral History Collection.

Private First Class Kenneth E. Jackson
Silver Star Citation

The President of the United States of America takes pride in presenting the Silver Star (Posthumously) to Private First Class Kenneth E. Jackson (MCSN: 2441265), United States Marine Corps, for conspicuous gallantry and intrepidity in action while serving as a Rifleman with Company G, Second Battalion, Ninth Marines, THIRD Marine Division (Rein.), FMF, in connection with combat operations against the enemy in the Republic of Vietnam. On 5 February 1969, Private First Class Jackson was in the point position during a platoon-sized patrol seventeen miles south of the Vandegrift Combat Base when he alertly detected a concealed enemy soldier. Reacting instantly, he notified his squad leader, and while the Marines began maneuvering forward, the point element came under intense hostile fire. In the initial burst of enemy small arms fire, Private First Class Jackson was painfully wounded in both legs. Steadfastly remaining in his dangerously exposed position, he began delivering suppressive fire upon the hostile soldiers, enabling two of his companions to maneuver out of the hazardous area to more advantageous positions. While continuing his aggressive efforts against the enemy, Private First Class Jackson was mortally wounded by hostile small arms fire. His heroic actions inspired all who observed him and his timely warning alerted his companions to the presence of the concealed enemy. By his courage, aggressive fighting spirit and unwavering devotion to duty, Private First Class Jackson upheld the highest traditions of the Marine Corps and of the United States Naval Service. He gallantly gave his life for his country.

For those of 2d Platoon who peered over the rocks, they could see Jackson, Brower, Head, and Bernstein directly in front of them. Henderson could hear Corpsman Bernstein struggling to say that he could not breathe. Others shouted out the positions of enemy soldiers. To Henderson's left was Private First Class Wade H. Cupps, who already had a close encounter with a rifle grenade that singed his trousers and another that peppered his right leg with shrapnel. He was surveying the situation when a burst of rifle fire ended a momentary lull in the shooting. At that moment, he saw movement to his left and a Marine sprinting in the open toward the pinned-down group, firing his rifle from the hip. Cupps recognized the Marine as Lance Corporal Thomas P. Noonan, from 2d Platoon.[141]

Back at the wounded group, Brower struggled to restrain Corpsman Bernstein, who attempted to crawl up the hill toward safety. As Brower held down the senior corpsman, trying to keep everyone's head down and not draw enemy fire, Lance Corporal Noonan suddenly appeared, crouched behind a rock only meters away. "How many of you are hurt?" he calmly asked. Brower told Noonan they were all injured, but Bernstein was the worst. "All right, hang loose," Noonan said. "I'll get you out of here. I'll take 'Doc' Bernstein first and then come back for the others."[142] He leaped out from behind the rock and gathered up Corpsman Bernstein. From the vantage point of 2d Platoon, they watched as a rifle grenade exploded in the group while Noonan began pulling. The explosion staggered him, but he pulled himself back up, wrapped his arms around Bernstein and began dragging once again. Seeing what was happening, Cupps ran forward to the rock where Noonan earlier sought cover. Before he hit the ground, though, he watched as a burst of AK-47 rifle fire hit Noonan and Corpsman Bernstein, immediately killing both men.[143]

Witnessing a selfless act of heroism from one of their comrades, the Marines of Company G rallied. Second Lieutenant Albert J. Langford's 1st Platoon swung to the left of 2d

---

[141] 2dLt Jack Henderson, statement, in Noonan recommendation; and PFC Wade H. Cupps, statement, in Noonan recommendation.

[142] LCpl E. F. Brower, statement, in Noonan recommendation.

[143] PFC Wade H. Cupps, statement, in Noonan recommendation.

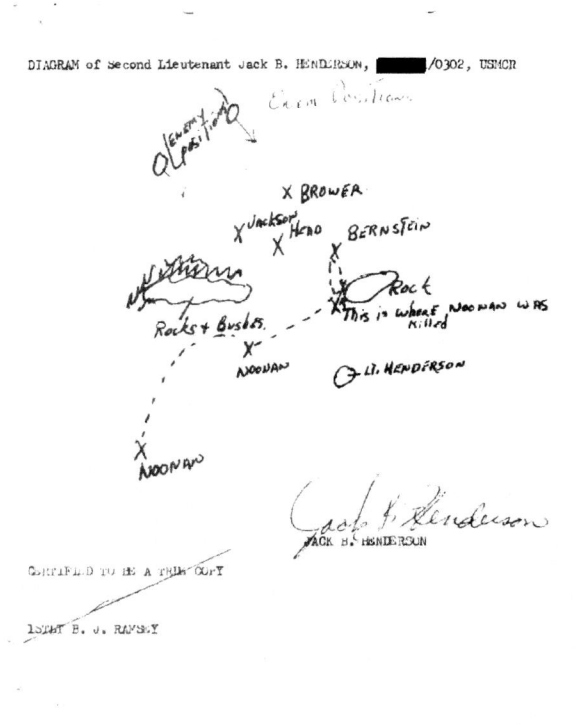

The map that 2dLt Jack B. Henderson, platoon commander of 2d Platoon, Company G, 2d Battalion, 9th Marines, drew for LCpl Thomas P. Noonan's Medal of Honor recommendation.
Archives Branch, Marine Corps History Division

Still high atop Co Ka Leuye, Company G needed to descend the mountain and seek refuge at Landing Zone Dallas in the valley below. Hitzelberger concluded that he and his Marines had only a half hour before a potential enemy attack. They used that time to destroy excess equipment and fashion stretchers to carry their dead and wounded. The company moved out at 1730, with a robust rear guard and a walking artillery barrage—a "steel wall," as Hitzelberger termed it—only 150 meters behind the column.[146] Half of the men carried stretchers or helped the wounded, which made traversing slick slopes in excess of 70 degrees time-consuming. Some difficult areas required a half hour for one stretcher to traverse and upward of 10 Marines to assist. For nine straight hours, Company G descended as quickly as they could. When darkness fell, artillery batteries dropped illumination rounds 50 meters in front of the column every two-and-a-half minutes to light the way, with each round offering 50 seconds of light.[147]

At 1400 the next day, Hitzelberger ordered his men to stop for sleep. They resumed after a few hours and reached the most arduous part of the journey in the evening, which required lowering their wounded and dead down a sheer rock cliff. At the bottom of the cliff, they linked up with a relief platoon that delivered much-needed medical supplies and rations, the first food the Marines had eaten in four days. Though the gradient became more manageable, the company still needed to find a suitable landing zone. It took another day-and-a-half to reach the Da Krong River, at the foot of Co Ka Leuye. Two CH-46 aircrews from HMM-161 displayed tremendous skill and courage, flying close to the river to see through the dense fog and avoid enemy antiaircraft fire. Landing on sandbars, they loaded the wounded and flew back up the river and on to Vandegrift Combat Base.[148]

Company G walked into Landing Zone Dallas on 9 February, ending a weeklong tribulation. They fared the worst of all 9th Marines units in Phase II, taking the most extreme terrain, being the most exposed for the longest, and fighting the largest engagement. For his selfless efforts to lay down suppressive fire during the firefight on 5 February, Jackson posthumously received the Silver Star. For his disregard for personal

Platoon and moved down a ravine to flank the enemy position.[144] With the enemy troops' attention now directed at the force flanking them, Hitzelberger ordered 2d and 3d Platoons to press the attack and push through and beyond the ambush site. The North Vietnamese force broke and ran, and Company G consolidated to take stock of their situation. They found two dead enemy soldiers and several blood trails leading into the jungle. Five Marines died in the firefight, and 18 were wounded.[145]

---

[144] 2dLt Albert J. Langford, interview with SSgt Willis S. Bernard Jr., 5–9 March 1969, Marine Corps Oral History Collection.
[145] Smith, *High Mobility and Standdown, 1969*, 35–36.
[146] Hitzelberger interview.
[147] Hitzelberger interview.
[148] Hitzelberger interview; and Smith, *High Mobility and Standdown, 1969*, 36.

Capt Daniel A. Hitzelberger (right) and GySgt Charles A. Baker (left).

safety in attempting to rescue Corpsman Bernstein, Noonan posthumously received the Medal of Honor.

The bad weather that began on 31 January blunted the momentum 9th Marines built in Phases I and II. For 10 days, Colonel Barrow and his staff waited for conditions to improve so they could resume the drive beyond Phase Line Red and into the southern Da Krong Valley. In that time, the enemy reinforced and improved their defenses.[149]

PHASE III

The drizzle slackened and the fog lifted on 9 February. The lost days meant a loss of momentum, which affected 9th Marines in a variety of ways. First, the regiment had to regain its jump-off position, which required sweeping operations around the fire support bases once again and a rearrangement of battalions. These changes necessitated an altered schedule. Barrow and his staff planned for the battalions to attack simultaneously abreast of one another when Phase III began. Now they settled for 3d Battalion launching on 11 February and 1st and 2d Battalions following the next day. Finally, the loss of momentum allowed the enemy to prepare a stubborn defense, something that they could not do during Phase II, when 9th Marines established fire support bases at an astonishing rate.[150] North Vietnamese commanders could deduce where the assault was

---

[149] Barrow to Davis report.

[150] Barrow interview.

Lance Corporal Thomas P. Noonan

Defense Department (Marine Corps) A700695

Thomas Patrick Noonan was born in Brooklyn, New York, on 18 November 1943. Following graduation from Grover Cleveland High School in Ridgewood, New York, in June 1961, he briefly studied at seminary before attending Hunter College in the Bronx, where he earned a bachelor's degree in physical education in June 1966.

While Noonan went to college, his close friend since kindergarten, Robert E. O'Malley, enlisted in the Marine Corps and fought as a squad leader with Company I, 3d Battalion, 3d Marines, in Vietnam.[1] During Operation Starlite, Corporal O'Malley led his squad in an assault against an entrenched enemy force on 18 August 1965. Sprinting across an open rice paddy, he jumped into the enemy trench line and killed eight People's Liberation Armed Forces fighters with his rifle and grenades before regrouping his squad and assisting another Marine unit that was taking casualties. After an officer ordered him to evacuate, O'Malley led his squad to helicopters and waited until all of his men were aboard before leaving the battlefield. For his actions, O'Malley became the first Marine Medal of Honor recipient of the Vietnam War.

His close friend's actions inspired Noonan to enlist in the U.S. Marine Corps Reserve in Brooklyn the day after Christmas, 1967. A month later, he discharged from the reserves to enlist in the regular Marine Corps and completed recruit training with 3d Recruit Training Battalion at Marine Corps Recruit Depot, Parris Island, South Carolina, in April 1968. After completing Individual Combat Training with 1st Battalion, 1st Infantry Training Regiment, at Camp Lejeune, North Carolina, Noonan deployed to the Republic of Vietnam in July 1968, where he was a motorman with Headquarters and Service Company, 2d Battalion, 27th Marines. One month later, he transferred to Company G, 2d Battalion, 9th Marines, and saw combat as a rifleman. On 1 January 1969, he received a promotion to lance corporal. For his attempt to save a wounded comrade on Hill 1175 during Operation Dewey Canyon, on 5 February 1969, Noonan posthumously received the Medal of Honor, marking the rare occurrence of two childhood friends both receiving the nation's highest military decoration for valor.

Thomas Noonan is buried in Calvary Cemetery, in the Sunnyside neighborhood of Queens, New York. His name appears on panel 33W, line 67, of the Vietnam Veterans Memorial in Washington, DC.[2]

---

[1] Denis Hamill, "A Day to Remember Childhood Pals from Queens Who Earned Medal of Honor," *New York Daily News*, 26 May 2012.

[2] Noonan recommendation.

A McDonnell Douglas F-4 Phantom II fighter-bomber provides close-air support to Marines.

heading, given that all indications pointed to an attack on Base Area 611 and the roads connecting it. With the Laotian border presenting a barricade of sorts for the Marines, maneuver space was more limited than it had been in their leapfrogging of hilltops down the Da Krong Valley. Each battalion had a zone of action five kilometers wide, and their objectives were an average of eight kilometers in front of them.[151] Knowing this, the enemy emplaced and mounted small harassing attacks on the troops who pulled back around the fire support bases during the inclement weather. Mostly, though, they waited for the Marines to come to them.[152]

Several companies obliged the NVA units on 9 February, when the Marines swept around their positions in preparation for crossing Phase Line Red. Though the enemy was dug in, the Marines could once again rely on close-air support as well as casualty evacuation and resupply helicopters. Sporadic firefights began at 0915 on 9 February when Company F, 2d Battalion, 9th Marines, walked into an ambush. The company returned fire, and the squad of enemy soldiers broke contact and fled southwest. The process repeated itself several times

---

[151] Davis, "Dewey Canyon," 37.

[152] Davis, "Dewey Canyon," 37.

that morning and early afternoon, with rifle companies engaging small groups of enemy troops, usually resulting in wounded Marines, dead North Vietnamese troops, and the enemy units fleeing southwest. Throughout the afternoon, the Marines discovered shelters, bunkers, and small arms and ammunition with increasing frequency.[153] The largest find that illustrated the enemy's supply nexus was a well-managed road, 10-meters wide with 4-meters clearance and drainage ditches.[154]

On 10 February, NVA units screening the main force were active against Marines pushing closer to Phase Line Red. Most contact during the morning for 9th Marines was in 3d Battalion's area, on the regiment's eastern flank, but 2d Battalion ran into enemy resistance by early afternoon on the western flank. At the same time, the remaining elements of 9th Marines not yet in the fight boarded helicopters at Vandegrift Combat Base, where they had been performing security. Companies B and C from Lieutenant Colonel Smith's 1st Battalion were bound for Fire Support Base Erskine, the southernmost fire support base that Company C, 3d Engineer Battalion, had not yet finished due to the poor weather.[155] Companies A and D, 1st Battalion, both took part in the initial phases of Operation Dewey Canyon and now waited to rejoin the battalion. First Lieutenant Fox's Company A spent much of the previous three weeks on Fire Support Base Shiloh, which they secured during Phase I. In the early hours of the day they were to rejoin the battalion, Fox and his Marines awoke to artillery rounds falling within their position. Hearing the report of guns in the distance, Fox quickly realized it was friendly fire and shouted over the radio to cease fire. The rounds landed in one of his company's machine-gun positions. In the darkness, the Marines found two team members dead in their dugout where they were sleeping. It was not until the sun came up that they found the gunner, dead. An investigation later learned that the deaths were the result of a howitzer crew from Battery F, 2d Battalion, 12th Marines, entering the wrong data when firing a harassment and interdiction mission two kilometers north.[156] After the friendly fire incident, Company A rejoined 1st Battalion and Fire Support Base Shiloh closed.[157]

The long-delayed Phase III of Operation Dewey Canyon commenced at dawn on 11 February, when Lieutenant Colonel Laine's 3d Battalion waded across the knee-deep Da Krong River. Colonel Barrow's plan was to drive to the Laotian border and sever the portions of Route 922 in the Republic of Vietnam. It meant his battalions would have to attack uphill, as the border was on a common ridgeline that ran east-west and took a sharp north-south turn, creating a fishhook. To protect the regiment's assault, Barrow ordered 3d Battalion, positioned on the regiment's left, to take the high ground. It would do so without Company L, which was performing security at Fire Support Base Cunningham. Tiger Mountain dominated the area, overlooking the border and offering a natural barrier between the Da Krong and A Shau Valleys. Five kilometers due south was Tam Boi, an equally important piece of terrain, where Routes 922 and 548 intersected.[158] With possession of these hills, 9th Marines could protect Phase III operations with artillery coverage.

The companies of 3d Battalion, 9th Marines, were to climb out of the river valley on the backs of two ridgelines 2,000 meters apart and assault Tiger Mountain.[159] The 1st and 2d Battalions, 9th Marines, were to attack objectives astride the Laotian border. In the middle was Lieutenant Colonel Smith's 1st Battalion, which would attack much like 3d Battalion. Upon reaching the Da Krong River, the battalion column was to split, with two companies advancing abreast along parallel ridgelines running south and two companies in trace in case of contact. The battalion would converge at Regimental Objective 2, gaining control of Routes 922 and 548. Farther west, on the regiment's right flank, Lieutenant Colonel Fox's 2d Battalion, 9th Marines, was to remain mostly in the valley floor but mount secondary attacks on the ridges that abutted

---

[153] See "Sequential Listing of Significant Events," in 9th Marines ComdC, 1 February to 28 February 1969, HD Archive, III-6.
[154] See "Sequential Listing of Significant Events," in 2d Battalion, 9th Marines ComdC, 1 February to 28 February 1969, HD Archive, III-2.
[155] Company C to 3d Engineer Battalion report.
[156] The command chronology reports that there were two short rounds, though Fox argues it was four and an inadvertent targeting. See 9th Marines ComdC, 1 February to 28 February 1969; and Col Wesley L. Fox, USMC (Ret) *Marine Rifleman: Forty-Three Years in the Corps* (Washington, DC: Potomac Books, 2002), 243–44.
[157] Barrow to Davis report.
[158] Davis, "Dewey Canyon," 37.
[159] Laine interview.

TACTICAL SITUATION, PHASE III

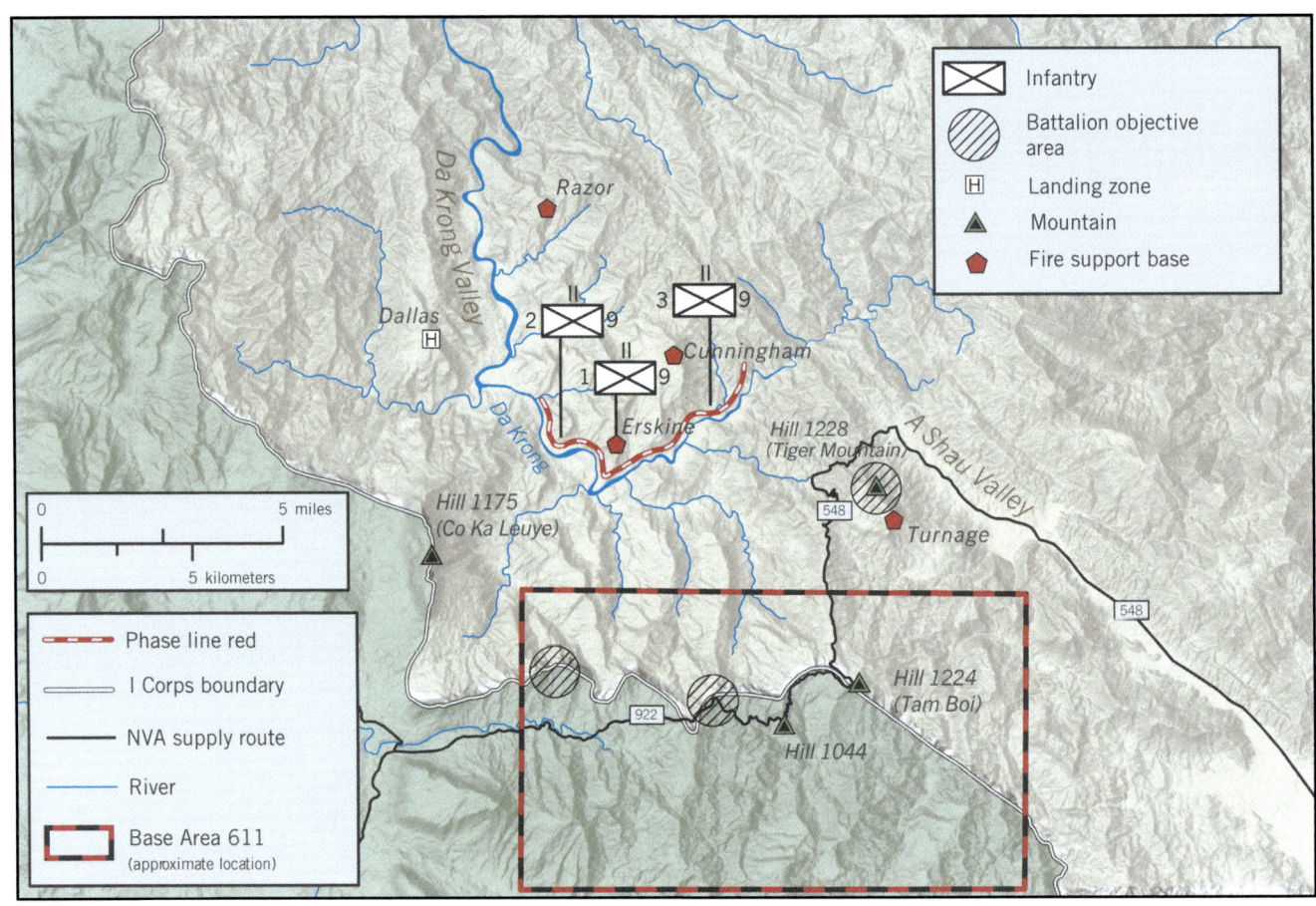

Map courtesy of Pete McPhail, adapted by MCUP

the boundary with 1st Battalion, 9th Marines.[160] Aiding each rifle company were helicopter support teams, engineers, Kit Carson scouts, and dogs from the 3d Military Police Battalion's Scout Dog Platoon.[161]

CROSSING THE DA KRONG RIVER

The experience of Captain Thomas F. Hinkle's Company M, 3d Battalion, 9th Marines, foretold what the rest of the regiment would find when they crossed Phase Line Red. After wading across the Da Krong River on the morning of 11 February and climbing the hills on the southern bank, the company's lead platoon confronted enemy soldiers in bunkers. As near as the Marines could tell, NVA troops hastily built the positions during the rain-induced lull in operations.[162] Captain Hinkle dispatched his 1st Platoon to assist, but it was unable to consolidate with 2d Platoon. Throughout the night, the enemy force used mortar-supported probes against the Marines' two perimeters with 20-man groups and fired on casualty evacuation helicopters attempting to carry out the wounded. The Marines employed much of their supporting arms against the enemy positions, with 60mm and 81mm mortars as well as artillery from Fire Support Base Cunningham. UH-1Es stayed until night fell, when Douglas AC-47 Spooky gunships took over for almost six hours. Pilots banked

---

[160] Davis, "Dewey Canyon," 37.
[161] Capt Edward F. Riley, interview with Sgt Marshall Neal Jr., 19–28 April 1969, Marine Corps Oral History Collection, hereafter Riley interview. Kit Carson scouts were NVA soldiers or Viet Cong guerrillas who defected and assisted American and ARVN units in the field.

[162] 2dLt Dallas M. Hyde, interview with SSgt Willis S. Bernard Jr., 18 March 1969, Marine Corps Oral History Collection, hereafter Hyde interview.

to portside so crews could fire three 7.62mm miniguns, emitting what appeared as one continuous orange flame from the aircraft to the ground. Below, 2d Platoon commander Second Lieutenant Willard P. Armes blew small chunks of C-4 explosive for much of the night to mark his perimeter for the AC-47 crew.[163] At 0500, 2d Platoon heard movement in front of their lines. An enemy force then stood up and advanced while screaming. The Marines rebuffed the attack before 1st Platoon finally linked up before dawn and helped evacuate the casualties. Enemy soldiers fired at and disabled one of the casualty evacuation helicopters when it arrived.[164] Their preferred tactic was to employ 82mm mortars only when helicopters were landing, using the noise and dust from the rotors to mask the mortar tube's location.[165] By daylight, what remained of the enemy force withdrew, leaving 18 dead and evidence of more casualties. Once Company M moved out, it immediately made contact again. They waged a running fight for the next three hours, raising their casualty count in less than a day to 9 dead and 14 wounded.[166]

At 2200 on 11 February, a NVA company attacked Companies A and C and the 1st Battalion command group, who were on a ridge extending from Fire Support Base Erskine, one kilometer south. The Marines surmised the enemy force aimed to attack the fire support base but ran into the two companies instead. On the parallel ridgeline, several sappers attempted to penetrate the perimeter of Company D, 1st Battalion, 9th Marines. Anxious, the entire battalion opened up with a deafening barrage of small arms, machine guns, grenades, and M-79 grenade launchers.[167] In the darkness, five batteries from 2d Battalion, 12th Marines, laid fire on the attackers and forced a retreat. The enemy left behind 29 dead as well as weapons, equipment, and explosives. The two-hour firefight resulted in 2 Marines killed and 11 wounded.[168]

Lieutenant Colonel Smith's 1st Battalion, 9th Marines, nonetheless stepped off the next morning at 0730 on a humid and hot 12 February, with Company B securing the river crossing and Company C in the lead.[169] Farther west, on the regiment's right flank, Lieutenant Colonel Fox's 2d Battalion, 9th Marines, ran into less resistance than the regiment's 1st and 3d Battalions but still had a day bookended with ambushes that resulted in two Marines killed and six wounded.[170]

By the end of 12 February, the companies corroborated what they thought would be true—that the enemy used the lull in fighting to prepare fighting holes and bunkers and intended to defend the border area. Through prisoners and captured documents, the Marines also confirmed that they were now facing infantry units, mostly the People's Army of Vietnam's *6th Regiment*. Different than the small groups of support troops the Marines previously encountered, the enemy south of the Da Krong River was fresh, well trained and equipped, and highly motivated.[171] Along most avenues of approach, particularly in the eastern area of operations, the enemy employed four-man squads with light machine guns and rocket-propelled grenades every 100–200 meters. The squads maintained contact while falling back to the next prepared positions. Snipers tied in trees fired on the advancing Marines. Units repeated the process until reaching high ground, which they defended for as long as possible.[172]

Prior experience in mountain operations allowed 9th Marines to overcome the stiff resistance. With a company in trace for every company attacking, the battalions maintained tempo with a concept built around mutual support. While the company in the attack deployed to overcome enemy forces on contact, the supporting unit fired mortars and cleared a land-

---

[163] 2dLt Willard P. Armes, interview (no interviewer or date), Marine Corps Oral History Collection.
[164] 9th Marines ComdC, 1 February to 28 February 1969.
[165] Hitzelberger interview.
[166] Capt Thomas M. Hinkle, interview with SSgt Willis S. Bernard Jr., 21 March 1969, Marine Corps Oral History Collection, hereafter Hinkle interview. See also 3d Battalion, 9th Marines ComdC, 1 February to 28 February 1969, HD Archive, 24-26.
[167] Fox, *Marine Rifleman*, 244.

[168] See "Sequential Listing of Significant Events," in 1st Battalion, 9th Marines ComdC, 1 February to 28 February 1969, HD Archive, 9; Smith, *High Mobility and Standdown, 1969*, 38; and LtCol George W. Smith, interview with Sgt Marshall Neal Jr., 23 April 1969, Marine Corps Oral History Collection, hereafter Smith interview.
[169] 1st Battalion, 9th Marines ComdC, 1 February to 28 February 1969.
[170] 2d Battalion, 9th Marines ComdC, 1 February to 28 February 1969.
[171] Barrow to Davis report; and 1stLt Robert T. Rohweller, interview with Sgt Marshall Neal Jr., 14 April 1969, Marine Corps Oral History Collection, hereafter Rohweller interview.
[172] Maj William P. Negron, interview with Sgt Marshall Neal Jr., 6 April 1969, Marine Corps Oral History Collection, hereafter Negron interview; and Rohweller interview.

Second Lieutenant Willard P. Armes
Silver Star Citation

The President of the United States of America takes pleasure in presenting the Silver Star to Second Lieutenant Willard Paul Armes (MCSN: 0-106503), United States Marine Corps, for conspicuous gallantry and intrepidity in action while serving as a Platoon Commander with Company M, Third Battalion, Ninth Marines, THIRD Marine Division (Rein.), FMF, in connection with combat operations against the enemy in the Republic of Vietnam. On the evening of 10 February 1969, Second Lieutenant Armes' platoon was conducting a search and destroy operation near the Da Krong River when the Marines came under intense hostile small arms and automatic weapons fire. Reacting instantly, Second Lieutenant Armes rushed forward to the lead element of his platoon and quickly deployed his men, directing his machine gunners to establish a base of fire which enabled him to maneuver his platoon into more defensible positions. Lacking a platoon sergeant and other senior noncommissioned officers, he constantly moved among his men throughout the entire fire fight, directing their fire, supervising and assisting with the evacuation of the wounded, and maneuvering his unit toward its objective. Repeatedly exposing himself to enemy fire, he fearlessly encouraged his men to move forward until they secured their objective late the following morning, accounting for twenty-five hostile soldiers killed during the entire battle. His heroic and timely actions inspired all who observed him and were responsible for the successful completion of his platoon's mission. By his courage, bold leadership and unwavering devotion to duty in the face of great personal danger, Second Lieutenant Armes upheld the highest traditions of the Marine Corps and of the United States Naval Service.

---

ing zone in the jungle to bring in resupplies and evacuate the wounded. The company in support then passed through the reorganizing lead element to continue the attack.[173] The concept offered 9th Marines the opportunity to overcome rugged terrain and an emplaced enemy stubbornly defending its territory, but it was manpower intensive, requiring a full battalion with companies operating in close proximity to one another. As the operation dragged on, these maneuvers became more difficult.[174]

ATTACK ON FIRE SUPPORT BASE CUNNINGHAM

In the first week of Phase III, 9th Marines found that the enemy preferred to attack with reinforced platoons or companies. As throughout the rest of I Corps, NVA units attacked in the early morning hours, using darkness to counterbalance superior American supporting arms. This compelled some units to use what Lieutenant Colonel Smith and 1st Battalion termed "violent response," one solid minute of fire from all organic weaponry directed at the probe.[175] The enemy employed such an attack on 17 February, when a sapper platoon with a reinforced company assaulted Fire Support Base Cunningham after the Democratic Republic of Vietnam unilaterally declared a cease-fire for the Tet holiday. A few days prior, General Abrams sent an alert order notifying units that a 24-hour cessation of operations would begin at 1800 on 16 February. Learning from the Tet holiday the previous three years, when enemy troops attacked during a truce to varying degrees, Abrams ordered all units to maintain a high state of readiness and respond to any NVA or Viet Cong aggression with impunity.[176]

Fire Support Base Cunningham was an enticing target for enemy forces. Destroying it would eliminate one of the two southernmost bases that provided the crucial artillery fan

---

[173] Smith interview.
[174] Davis, "Dewey Canyon," 38.
[175] Smith interview.
[176] Gen Creighton Abrams to all units, order, 15 February 1969, attached to 1st Battalion, 9th Marines ComdC, 1 February to 28 February 1969.

Marines from Company H, 2d Battalion, 9th Marines, wade across the Da Krong River to begin Phase III of Operation Dewey Canyon.
Defense Department (Marine Corps) A800480

to the companies pushing into Base Area 611. It is unclear if enemy forces were aware of two other high-profile targets on Fire Support Base Cunningham, however. The first was Colonel Barrow's forward headquarters, which he moved from Fire Support Base Razor on 11 February. If the sappers were successful, wreaking havoc at the Marine position could disrupt command and control at a crucial moment in the operation. The second was a valuable communications terminal for processing intelligence. Destroying it could severely impede 9th Marines' ability to see what was in front of it and to coordinate the battalions. Crucially, the terminal required generators to power a number of machines, including a KW-7 electronic cipher for message encryption. The sound of the generators operating at night was a homing beacon for the enemy and made it difficult for First Lieutenant Raymond C. Benfatti's Company L, 3d Battalion, 9th Marines, which was providing base security, to hear the approaching force. Complicating matters for First Lieutenant Benfatti's Marines was heavy fog and the sporadic firing of the artillery batteries behind them.[177]

At 0345 on 17 February in the 2d Platoon command post, Second Lieutenant Milton J. Teixeira's radioman woke him to say that enemy troops just hit the platoon's listening post with small arms fire and grenades. At that moment, sappers entered the wire and engaged Teixeira's 1st Squad. Recoilless rifles and rocket-propelled grenades raked the platoon while mortar rounds exploded. By the time 1st Squad fired back, sap-

---

[177] 3d Battalion, 9th Marines ComdC, 1 February to 28 February 1969. See also Maj James S. Rayburn, "Direct Support during Operation DEWEY CANYON (U)," *Cryptologic Spectrum* 11, no. 3 (Summer 1981): 19.

pers were already on the bunkers and a new wave hit 3d Squad. The night was pitch-black, with no moonlight to illuminate the enemy storming 2d Platoon's wire. Defenders only saw shapes, not knowing if they were friendly or enemy. Marines threw tripflares in front of their positions in the hopes that they would ignite brush fires and backlight the attackers. The biggest problem the platoon faced, though, was fighting while evacuating the wounded and resupplying ammunition. Engineers and troops from Headquarters and Service Company came down the hill from Fire Support Base Cunningham to help with casualties while 1st Platoon, which was not being probed, ran ammunition to the other platoon's position.[178]

At the company command post, shrapnel from an exploding rocket-propelled grenade wounded Benfatti. He refused medical attention and rallied his Marines, directing their fire at the most critical points in the perimeter while enemy mortar rounds dropped on the company's position. The sappers who got through rushed to bunkers where Marines slept or worked and tossed in satchel charges and concussion grenades. One of their targets was 2d Battalion, 12th Marines' fire direction center. Several blasts rocked the bunker, scattering radios and other equipment and knocking the watch officer unconscious. Though the sappers succeeded in disrupting the fire direction center, the artillerymen at the batteries responded quickly and decentralized technical fire direction to continue without interruption, all while enemy mortarmen targeted gun emplacements. One of the 2d Battalion, 12th Marines' howitzers in Battery E fell victim to a mortar round, but Battery D, 3d Provisional Battery, and Battery W continued firing.[179]

Company L, 3d Battalion, 9th Marines, repulsed what remained of the sapper platoon. Marines farther within the base killed those who got through the wire. The defenders were astounded to watch sappers dress their wounds after being shot and continue fighting. They later suspected that some enemy fighters were under the influence of drugs.[180] Before the NVA company attacked, Benfatti organized a reaction force, oversaw the evacuation of his wounded, and directed

Douglas A. Yeager Collection, Archives Branch, Marine Corps History Division

A typical 12th Marines 105mm battery position during high mobility operations.

where in the line Marines should go. The artillerymen also reestablished control of the situation, restoring the fire direction center within 30 minutes. They unleashed direct fire on the larger force swarming Fire Support Base Cunningham from three sides. Using beehive rounds, designed to disrupt massed infantry assaults with thousands of steel darts called flechettes, the artillery pieces fired over Company L. Howitzers on Fire Support Bases Erskine and Lightning shelled the massed enemy troops, with some using controlled fragmentation munition rounds, each of which contained dozens of golf ball-size bomblets.[181]

For the next three hours of intense fighting in the dark, enemy troops surged up the hill and the Marines beat them back. By 0700, the enemy withdrew. To destroy what remained of the attacking force, the artillery batteries and fixed-wing aircraft targeted suspected escape routes and assembly areas. In all, 2d Battalion, 12th Marines, expended 3,270 rounds in four hours. Benfatti, who stayed with his company throughout the night and continually exposed himself to hostile fire, supervised the casualty evacuation of his men and refused attention to his own wounds until corpsmen tended to everyone

---

[178] 2dLt Milton J. Teixeira, interview with SSgt Willis S. Bernard Jr., 20 March 1969, Marine Corps Oral History Collection, hereafter Teixeira interview.
[179] 3d Battalion, 9th Marines ComdC, 1 February to 28 February 1969. See also 2d Battalion, 12th Marines ComdC, 1 February to 28 February 1969.
[180] Teixeira interview.

[181] Maj Joseph B. Knotts, interview with SSgt Willis S. Bernard Jr., 8 April 1969, Marine Corps Oral History Collection; and Teixeira interview.

Hospitalman Third Class Mack H. Wilhelm

Navy Cross Citation

The President of the United States of America takes pride in presenting the Navy Cross (Posthumously) to Hospitalman Third Class Mack H. Wilhelm (NSN: B-713921), United States Navy, for extraordinary heroism on 19 February 1969 as a Corpsman serving with Company D, First Battalion, Ninth Marines, THIRD Marine Division (Reinforced), Fleet Marine Force, in connection with combat operations against the enemy in the Republic of Vietnam. When his company came under a heavy volume of fire from an enemy force occupying a well-concealed bunker complex at the crest of a hill in the northern section of the I Corps Tactical Zone, Petty Officer Wilhelm observed a seriously wounded Marine lying dangerously exposed to the intense hostile fire, and quickly raced across the fire-swept terrain to the side of the casualty. Although Petty Officer Wilhelm was painfully wounded in the shoulder, he skillfully administered emergency first aid to his companion, picked him up and, shielding him with his own body, commenced to carry him to a sheltered position. Once again wounded, this time in the leg, Petty Officer Wilhelm nonetheless managed to evacuate his patient to a relatively safe location. He then returned through the hail of fire to the side of another critically wounded Marine and was in the process of examining the casualty when he, himself, was mortally wounded by a burst of enemy rifle fire. By his daring initiative, outstanding courage, and selfless dedication, Petty Officer Wilhelm was directly instrumental in saving the life of a fellow serviceman. His heroic and determined efforts were in keeping with the highest traditions of the United States Naval Service.

Jonathan F. Abel Collection, Archives Branch, Marine Corps History Division

HM3 Mack H. Wilhelm treats a Marine wounded after an NVA attack on Company D, 1st Battalion, 9th Marines' position. Days later, he died in the action for which he posthumously received the Navy Cross.

---

else. For his actions, he received the Silver Star. The Marines suffered 4 killed and 46 wounded but could have taken more casualties had it not been for faulty enemy weapons. Strewn about the positions were homemade grenades that failed to detonate, seven of which Teixeira found in one fighting position.[182] Thirteen sappers made it inside the perimeter before being killed, and 27 died in front of the defenses.

With the regiment picking up momentum in Phase III, 9th Marines needed its full strength south of the Da Krong River. Elements of 2d Battalion, 3d Marines joined the operation to ensure that each battalion of 9th Marines would have four rifle companies in the attack. On 18 February, Company G, 2d Battalion, 3d Marines, relieved Company L, 3d Battalion, 9th Marines, at Fire Support Base Cunningham.[183] Company L then helilifted 6.5 kilometers southeast, linking up with Company I, 3d Battalion, 9th Marines, at the base of Tiger Mountain in preparation for taking the dominant terrain feature between the Da Krong and A Shau Valleys.[184]

As 9th Marines advanced, signals intelligence became increasingly important. Colonel Barrow's headquarters had been listening to enemy radio traffic since the first week of

---

[182] Teixeira interview.
[183] Barrow interview.
[184] 3d Battalion, 9th Marines ComdC, 1 February to 28 February 1969.

the operation. On 25 January, a small team consisting of a gunnery sergeant serving as team chief (later an officer), two Marines, and an ARVN interpreter left Fire Support Base Razor to operate independently in the field. Outfitted lightly enough that they could move quickly when necessary, the direct support team from the 1st Radio Battalion used wire strung between trees for an antenna and an AN/PRR-15 radio for voice collection. The three Marines kept a 24-hour monitor of enemy communication networks, notifying the ARVN member of the team when enemy troops communicated to intercept the traffic. After decrypting the messages, the team chief evaluated what he deemed important and passed the information to the regimental intelligence officer. Simplicity made the operation mobile but also kept it vulnerable and isolated from the battalions.[185]

Augmenting the signals intelligence effort were teams from a provisional company of the 3d Reconnaissance Battalion. An important component of Major General Davis's high mobility concept, reconnaissance teams funneled information to Barrow's headquarters about trails and enemy movements, allowing 9th Marines to pinpoint objectives and maintain tempo. At any given time during Operation Dewey Canyon, there were 9 teams operating. By 5 March, there would be 45 in the area of operations.[186]

The direct support team made its largest impact during Phase III—particularly during the most intense period of combat for the entirety of the operation, 18–22 February. During those four days, the battlespace shrank, with 9th Marines racing to the border to destroy enemy units before they escaped, and North Vietnamese forces struggling to keep open the main supply routes to their border sanctuary.[187] Shortly after entering the field in Phase I, the direct support team identified enemy transportation and artillery communication networks. The intelligence helped a frustrated Barrow understand what assets the enemy was maneuvering and which 9th Marines' companies NVA forces were targeting. His exasperation was due to the American and ARVN inability to sever Route 922. Despite artillery fire, Arclight strikes from U.S. Air Force Boeing B-52 Stratofortress bombers, and Navy and Marine fixed-wing attacks, the enemy continued using the road. This created, in Barrow's mind, a "pretty unacceptable situation" that "cried out for some sort of action to put a stop to it."[188] He ordered artillery batteries to step up their road interdiction. The increase yielded results by 17 February, when the direct support team listened to NVA commanders complaining about difficulties moving troops and equipment from Laos into the Republic of Vietnam.[189]

122MM FIELD GUNS

Unsafe roads and a lack of vehicles prevented enemy artillery units from displacing their 122mm field guns.[190] On 19 February, 1st Battalion, 9th Marines, pushed forward, attempting to seize Regimental Objective 2 on the border before enemy units could regroup. When Company C under Captain John A. Kelly advanced on Hill 781, it ran into a robust bunker complex. NVA soldiers in well-camouflaged bunkers waited until Marines came within one meter before opening fire. The stiff defense differed from other parts of the area of operations, where enemy troops defended a ridgeline briefly before pulling back to the next line of prepared defenses. When enemy troops stood and fought, their bunkers proved somewhat impervious to anything but fixed-wing ordnance. Aircraft dropped what pilots termed "snake and nape," or 250- and 500-pound "snake eye" bombs with 500-pound M-47 napalm canisters.[191] There was, however, little room to use close-air support or maneuver on the narrow ridgeline. Captain Kelly had little alternative but to attack through the position the next morning. That night, the company heard diesel engines and track noise to the south, below their position on Hill 781. Thinking the commotion might be armor, Companies A and C prepared to fight tanks.[192]

Kelly's lead platoon, with Second Lieutenant Archie J. Biggers commanding, resumed the attack on the bunker complex at 0700 on 20 February. After assaulting through the

---

[185] Rayburn, "Direct Support during Operation DEWEY CANYON," 16–17.
[186] Barrow to Davis report.
[187] Capt Wesley L. Fox, interview with Staff Sergeant Willis S. Bernard Jr., 19–28 April 1969, Marine Corps Oral History Collection, hereafter Fox interview.

[188] Barrow interview.
[189] Rayburn, "Direct Support during Operation DEWEY CANYON," 18.
[190] Rayburn, "Direct Support during Operation DEWEY CANYON," 18.
[191] Negron interview.
[192] Capt John A. Kelly, interview with SSgt Willis S. Bernard Jr., 19–28 April 1969, Marine Corps Oral History Collection, hereafter Kelly interview.

A CH-46 from HMM-161 unloads Marines from Company L, 3d Battalion, 9th Marines, at the base of Tiger Mountain, 18 February 1969. The airframe, Bureau Number 154014, remained in the Marine Corps' inventory after Vietnam and was one of the last CH-46s the Service retired in 2015.

LOCATION OF 122MM FIELD GUNS

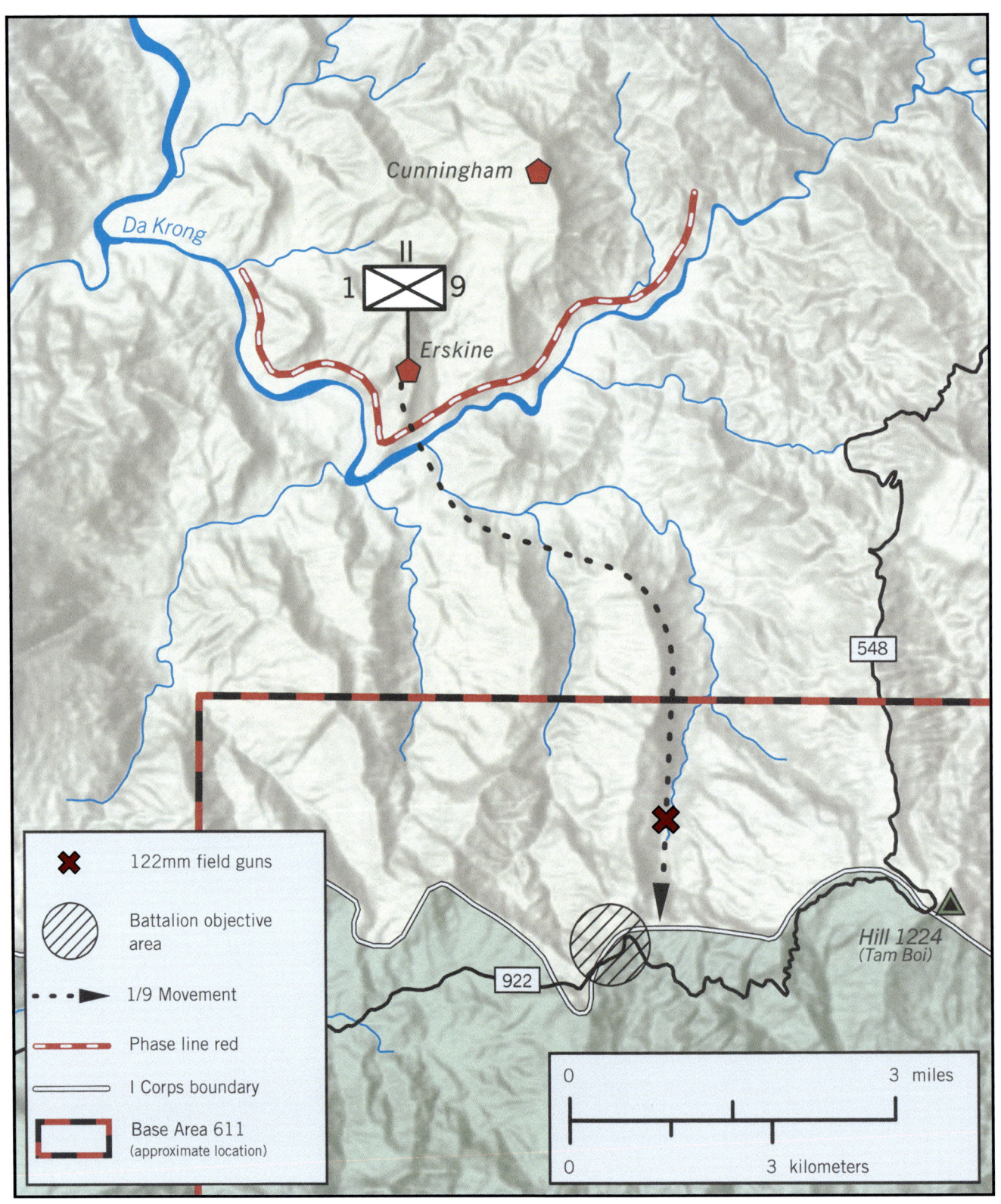

Map courtesy of Pete McPhail, adapted by MCUP

complex and toward a hill that Kelly declared an intermediate objective, the company broke out into a clearing that left them believing they were walking into Dante's *Inferno*. For two days, the Marines called in artillery and air strikes on the location, including 4,000 rounds from five batteries, bombs from attack aircraft, and an AC-47.[193] The ordnance left little standing. Company C then ran into more prepared positions and heavy resistance, gaining ground 30 meters at a time. Douglas A-4E Skyhawk light attack aircraft pilots from Marine Attack Squadron 211 (VMA-211) dropped napalm only 50 meters from the forward platoon's position, alternating fire with artillery batteries.[194] The company's ground attack eventually succeeded, killing 27 defenders; air strikes and artillery killed another 44 but taking the hill yielded an even greater prize.[195]

Along a narrow trail whose thick jungle canopy made it virtually invisible to aircraft, Biggers's platoon found two camouflaged 122mm field guns ready for transport. One was spiked, and an undetonated satchel charge lay beneath the second.[196] The tracked vehicle the company heard the night before was a Soviet AT-T heavy artillery prime mover, halfway up the hill from the guns and a burnt wreck because of a direct hit from a napalm canister. Company C later found nearby gun pits, complete with documents and firing sketches that revealed the company captured two of the artillery pieces that fired on Fire Support Base Cunningham on 2 February.[197] Kelly and his Marines deduced that the relentless artillery and air strikes that Barrow ordered convinced the enemy crews to flee to safety. The existence of four gun pits revealed that the battery commander was able to withdraw two guns but was forced to abandon the remaining two when the air strike destroyed the prime mover.[198] Higher headquarters judged the capture significant and ordered 1st Battalion, 9th Marines, to protect the guns until helicopters could evacuate them.[199]

When Company C paused to rest and take stock of their 4 dead and 24 wounded, First Lieutenant Fox's Company A, 1st Battalion, 9th Marines, pushed through to reach Battalion Objective Foxtrot, 1.5 kilometers from Regimental Objective 2 and the Laotian border. There they brushed aside an NVA platoon of service and support troops in successive squad defensive positions protecting a sizable depot.[200] Enemy resistance and the capture of the field guns convinced Lieutenant Colonel Smith to reconsolidate his battalion ahead of schedule, which gave him the ability to place a cordon around the guns while also pushing on to the objective.[201] It was no coincidence that Smith's 1st Battalion, 9th Marines, found the complex. Since Barrow knew the enemy was complaining about 2d Battalion, 12th Marines' emphasis on road interdiction, he was confident that the depot was lightly defended, and sent Smith's battalion forward to exploit the situation. Swiftly acting on such intelligence from the reconnaissance and direct support teams often had a compounding effect and generated more usable information. When 1st Battalion's companies advanced toward Route 922, the enemy artillery communication network erupted with chatter that revealed the existence of two more 122mm field guns as well as four 85mm pieces.[202]

During the next two days, 1st Battalion, 9th Marines, discovered the full extent of the supply area at Lang Ha-Bn located on the Laotian border.[203] Company A found caches of weapons, ammunition, and medical supplies in 100 reinforced bunkers. Company B discovered evidence of a well-developed hub that the enemy used to transfer materiel across the border, with extensive foot trails, communication equipment, and cooking and sleeping areas.[204] Given the sheer number of roads, camps, and fortifications Marine units were discovering, the question arose among the companies whether they were uncovering a network of installations or one vast complex that served the entire *People's Army of Vietnam 7th Front*.[205] Regardless, by taking the supply area at Lang Ha-Bn, 9th Marines finally reached Route 922. They could now sever a portion of the enemy's logistics chain and travel down the main supply route, seizing more depots as they went.

---

[193] Kelly interview; and After Action Report, VMA-211, February 1969, entries for bombing runs at 1032 and 1112 on 18 February 1969, HD Archive.
[194] See After Action Report, VMA-211, February 1969, entry for 0930 on 20 February 1969, HD Archive.
[195] See Barrow to Davis report, 10–11.
[196] Smith to Barrow report.
[197] Kelly interview; and Smith to Barrow report.
[198] Kelly interview.
[199] Kelly interview.

[200] Fox interview.
[201] Riley interview.
[202] Rayburn, "Direct Support during Operation DEWEY CANYON," 18.
[203] Fox, *Marine Rifleman*, 250–51.
[204] 1st Battalion, 9th Marines ComdC, 1 February to 28 February 1969.
[205] Smith to Barrow report.

D-74, 122mm Field Gun

Photo courtesy of Kater Miller

The 122mm field gun as it appears today at the National Museum of the Marine Corps.

One of the two 122mm field guns that Company C, 1st Battalion, 9th Marines, captured in the vicinity of Lang Ha-Bn resides today in the National Museum of the Marine Corps, in Triangle, Virginia. It can be seen in the Vietnam War Gallery. Its path to the museum involved Marine ingenuity, some scheming, and a two-star general.

Major General Raymond G. Davis, 3d Marine Division commander, took particular interest when 1st Battalion, 9th Marines, reported what they found on 20 February. The Soviet-made 122mm field guns were the largest artillery pieces American forces had captured in the war. Major General Davis ordered the guns secured and evacuated as soon as possible. It proved difficult, however, sourcing heavy-lift helicopters during a busy operation already off schedule because of persistent rain and low cloud cover. Lieutenant Colonel George W. Smith and his 1st Battalion remained in a defensive perimeter around the guns for one week, with the enemy harassing and probing most nights. Helicopters evacuated the first gun on 25 February 1969. The second followed the next day.[1]

Word about the guns spread quickly throughout Vietnam, and various headquarters clamored for securing one of the trophies. After the first gun went to Saigon and then disappeared off his radar, Davis went to great lengths to ensure that the second gun arrived where he wanted it. On 14 March, he wrote to Colonel John H. Magruder III, director of Marine Corps Museums. "I have a deal working to get one of our captured 122mm guns headed our way," he reported. There was "so much 'high level' interest in getting one for display," though, it was necessary to "get this one out quietly." Davis's plan was to have the gun "smuggled" through Okinawa aboard an Air Force Lockheed C-130 Hercules.[2]

Davis sent the gun to Da Nang for handover to USMACV but not before ordering a placard welded to the trail that read, "Captured 20 Feb 69 A Shau Charlie Co 1st Bn 9th Marines." Days later, Marines loaded the gun on a C-130 bound for Marine Corps Base Quantico. Accompanying the gun was Staff Sergeant Leonard L. Miller, who had orders to stay with the gun at all times. Miller took his orders so seriously that he remained inside the aircraft for two days in Alaska while the air crew repaired a mechanical failure, still dressed in his jungle fatigues. Though he contemplated changing

---

[1] Fox interview.
[2] MajGen Raymond Davis to Col John H. Magruder III, letter, 14 March 1969, in National Museum of the Marine Corps files, Heavy Ordnance Collection, accession number 1983.681.1, hereafter NMMC files.

Jonathan F. Abel Collection, Archives Branch, Marine Corps History Division

MajGen Raymond G. Davis inspects the 122mm field gun before it leaves Vietnam for Quantico, VA.

into his greens, Miller reasoned that he should not wear the wrinkled uniform until he could get it pressed. The gun—and Miller—arrived at Quantico on 21 March.[3]

Three days later, Colonel Magruder wrote to thank Davis for the lengths the general went to "bootleg the gun to us."[4] The museum temporarily displayed the gun at The Basic School and then moved it to a storage pad at Quantico. It remained there until 2004, when it moved to the Aberdeen Proving Ground in Maryland. Two years later, it returned to the Quantico area and was placed on display in the new National Museum of the Marine Corps, serving as the primary artifact in the High Mobility Operations section of the Vietnam exhibit. It remains there today.

---

[3] Col John H. Magruder III to MajGen Raymond Davis, letter, 24 March 1969, NMMC files, hereafter Magruder to Davis letter.
[4] Magruder to Davis letter.

COMPANY A AND THE BUNKER FIGHT

First Lieutenant Fox intended to do just that when he set off with his company across the border on 21 February. The morning was relatively quiet. His Marines entered Laos around 1030 and uncovered more caches. Company B went ahead, cutting Route 922, and eventually going 1,200 meters beyond the border, where it found 122mm ammunition and destroyed trucks.[206] At midday, a patrol from Company C, the unit securing the captured field guns, made heavy contact to their east. The appearance of NVA troops near the battalion's prizes made Lieutenant Colonel Smith, whose command post was nearby, believe that enemy units were making a bid to recapture the guns. He ordered Fox to send a platoon to assist Company C so the battalion could conduct a reconnaissance in force in a valley to the east and determine what the enemy was doing.[207] Fox sent Second Lieutenant William J. Christman III, whose 3d Platoon ran into an enemy force of indeterminate size shortly after arriving on the scene. Interpreting the heavy contact as confirmation that the guns and therefore Company C and Headquarters and Service Company were in jeopardy, Smith directed Fox to leave the border and reinforce Captain Kelly's position. When Fox arrived to extend Company C's line and thus the battalion's left flank, Second Lieutenant Christman and his platoon were still in contact. They remained pinned down until nightfall, when Christman retrieved the bodies of two Marines who were killed in front of the enemy position and reunited his platoon with the company.[208]

The next morning, Smith planned to destroy the enemy force threatening him. He instructed Company A to screen the battalion's left flank and search the area from the previous day's firefight.[209] If Fox did not meet resistance, he was to move two kilometers down the hill to a creek bed, where he would then turn right, search the valley floor, and finally return to the battalion perimeter around the captured guns. The first checkpoint at the creek bed was to serve a dual purpose. Bad weather prevented helicopter resupply and the battalion was low on water. If Fox arrived at the first checkpoint without incident, his orders were to summon a detail from Company C and Headquarters and Service Company and then move on once the battalion replenished its water supply.[210]

Christman, the officer most familiar with the area, guided Company A to where his platoon made contact the evening before. The triple-canopy foliage and a jungle floor choked with undergrowth and banana groves made everything look familiar, and Christman took the wrong fork on the trail. When he realized his mistake, the enemy position was at the company's right rear. Fox wagered that the opposing force pulled back overnight; otherwise, they would have attacked the vulnerable Marines. Company A pressed on to the creek bed and reached it at 1100, where an enemy squad inside a bunker killed 2d Platoon's point man and severely wounded another Marine before the lead squad overwhelmed the position.[211] Without further incident, Fox secured the first checkpoint and sent for the battalion water detail. While his company waited, they took the opportunity to eat rations and refill canteens. The detail arrived a half hour later, and with them came sporadic mortar fire from the enemy position the evening before that Christman inadvertently bypassed in the morning.[212] Thick foliage and branches provided a protective barrier between the Marines and the mortars, making the rounds more annoying than threatening. Equally bothersome was the machine-gun fire that came from the same enemy position. Rounds harmlessly ricocheted off trees. Fox determined that the enemy commander was provoking him into a fight, so he directed his platoons to move south, back up the hill, and toward the enemy position 200 meters away, setting in motion the largest pitched battle of Operation Dewey Canyon.[213]

When Fox's company approached the NVA positions around 1300, 1st Platoon could see that the enemy was in bunkers to their right flank. Fox ordered the 1st Platoon commander, Second Lieutenant George M. Malone Jr., to wheel to the right and attack west. He then ordered 3d Platoon to follow, which gave him two platoons in the attack and on

---

[206] 1st Battalion, 9th Marines ComdC, 1 February to 28 February 1969.
[207] Fox, *Marine Rifleman*, 251.
[208] Fox, *Marine Rifleman*, 251.
[209] Smith interview.
[210] Fox, *Marine Rifleman*, 251–52. Fox reports that the water detail included 15 men, but Charles R. Smith says it was 20. See Fox, *Marine Rifleman*, 252; and Smith, *High Mobility and Standdown, 1969*, 45.
[211] 2dLt James H. Davis, interview with SSgt Willis S. Bernard Jr., 19–28 April 1969, Marine Corps Oral History Collection.
[212] Fox writes that it was 82mm mortars in his memoir, but he says it was 61mm in his oral history interview from 1969. Charles R. Smith writes that it was 60mm. See Fox, *Marine Rifleman*, 252; Fox interview; and Smith, *High Mobility and Standdown, 1969*, 45.
[213] Fox, *Marine Rifleman*, 253.

Marines from Company A, 1st Battalion, 9th Marines, walk down an NVA trail along a ridgeline inside Base Area 611.

line, with 2d Platoon in trace.[214] His company was not at full strength. He left his mortar section with the battalion, figuring that they would prove useless in the thick jungle, and he was forced to send a squad with the water detail as an escort due to the sporadic firing. It quickly became obvious, however, that having a reduced force was the least of Fox's problems. As they approached the enemy position, the Marines melted into the dense vegetation, and soon Fox, who was following at the center of the assault, could only see individual rifleman as they used the jungle for cover. Now seeing the approaching Americans, the enemy increased the volume of fire. Small-arms fire and machine guns scythed through trees, branches, and leaves of the banana grove. Company A's assault remained orderly but slowed. At the moment before the attack stalled, an enemy rocket-propelled grenade landed only one meter behind Fox when he was providing a situation report to Lieutenant Colonel Smith on the radio, showering his left leg and shoulder with shrapnel. Recovering, he agreed with his battalion commander that Company D, 600–800 meters behind the enemy position, should assist.[215]

---

[214] Fox interview; and Barrow to Davis report.

[215] Fox, *Marine Rifleman*, 253; and Fox interview.

Marines went prone to avoid the withering fire. Fox did not realize that he was facing a reinforced company in a camouflaged, fortified, and mutually supporting bunker complex.[216] Behind him, on the other side of the creek, another enemy unit fired rocket-propelled grenades and mortars.[217] In front of him, one machine gun anchored the defense. Not only did it have a commanding view of Company A's approach, giving it almost an unobstructed field of fire, but the position was located across a ravine, out of reach of satchel charges or light antitank weapons.[218] Fox crawled forward to better understand the situation when a sniper in a tree fired at him but missed. Unloading half a magazine from his rifle, Fox killed the sniper. He returned to his command group and committed the depleted 2d Platoon, which had suffered two casualties from the morning and was less one squad due to the water detail escort. He ordered Second Lieutenant James H. Davis to attack the center of the enemy position, between the pinned down 1st and 3d Platoons. He determined a frontal attack was his only option: he could not break contact without risking more casualties and the dense jungle made a flanking maneuver too risky. When Fox relayed his plan to Second Lieutenant Davis, a mortar round landed within the command group, killing or wounding everyone except for the company executive officer, First Lieutenant Lee R. Herron. With Davis seriously wounded, Fox ordered First Lieutenant Herron to lead the attack.[219]

Almost immediately, the situation deteriorated further when machine-gun fire from the right flank wounded two Marines from 2d Platoon whom Fox recruited to replace his dead radiomen.[220] Minutes later, Fox received word from his three platoon sergeants. Second Lieutenants Malone and Christman were both seriously wounded and machine-gun fire killed First Lieutenant Herron. Meanwhile, a group of 20 North Vietnamese soldiers attempted to flank Company A, but a fireteam from the 3d Platoon, as well as a group of wounded Marines who Gunnery Sergeant Ronald G. Duerr organized, destroyed the flanking maneuver.[221] The low cloud cover that hung over the area meant Company A had no artillery or air support, and Company D was still likely too far away to be a factor. On the radio, Lieutenant Colonel Smith suggested the company pull back to saturate the area with artillery, but Fox judged that he already suffered so many casualties and was too close to the enemy that such an action would put his Marines at more risk.[222] Out of options, Fox, who had been wounded three separate times, was losing hope when two Marine OV-10 Broncos that Smith kept on station in case the weather improved appeared overhead. Seizing the opportunity, Fox ordered his platoon sergeants to mark their positions with smoke. He used the air frequency and call signs that battalion supplied to raise the aircraft and gave the pilots an azimuth and distance. The OV-10s' rocket and cannon runs were successful, knocking out the enemy machine gun that pinned down Company A.[223]

When the OV-10s left after their final attack, the shooting stopped. Fox's Marines, still prone on the ground, waited to see what the enemy would do next when Company D arrived just before dark and from behind the machine-gun position that the aircraft attacked. They, too, took the wrong trail, just as Christman had earlier that day.[224] Company D's unopposed march into the battle area meant that the opposing force was destroyed. Companies A and D spent the rest of the evening treating the wounded and collecting the dead, arriving back inside the battalion perimeter at 0400 the next morning.[225] They counted 105 NVA soldiers dead, and discovered that the unit, which was outfitted with new equipment and uniforms, was a veteran of previous campaigns, not a service and support unit like 1st Battalion had fought over the past few days. Fox's company suffered 12 killed and 72 wounded.[226] He lost all his platoon leaders: First Lieutenant Herron was killed, Second Lieutenant Christman later died of his wounds, and Second Lieutenants Malone and Davis were wounded. Company A's actions on 22 February represented one of the more decorated days in Marine Corps history, with seven Silver

---

[216] Smith interview.
[217] Fox interview.
[218] Fox interview.
[219] Fox, *Marine Rifleman*, 254; and Fox interview.
[220] Fox interview.
[221] Fox interview.
[222] Smith interview; and Fox interview.
[223] Fox, *Marine Rifleman*, 255.
[224] Fox interview.
[225] Riley interview.
[226] Smith, *High Mobility and Standdown, 1969*, 46; and 1st Battalion, 9th Marines ComdC, 1 February to 28 February 1969. While the 1st Battalion command chronology and Charles R. Smith say 11 killed in action, Fox writes that the company lost 12 men. The discrepancy is that 9th Marines' documents do not reflect 2dLt Christman dying from his wounds.

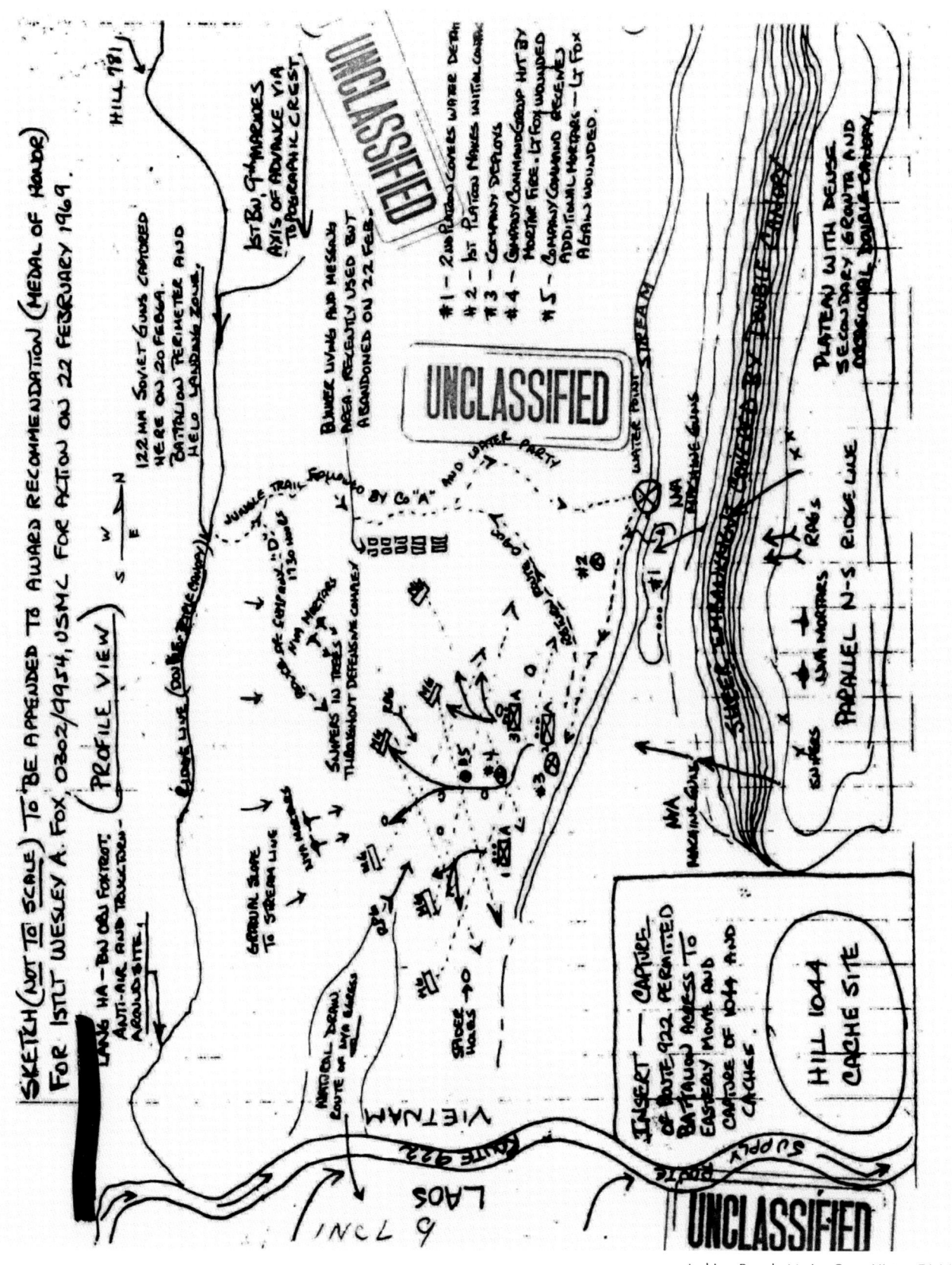

The hand-drawn map that accompanied the Medal of Honor recommendation for 1stLt Wesley L. Fox.

Defense Department (Marine Corps) 014140590

1stLt Wesley Fox, third from right, sitting on one of the two 122mm field guns that Company C discovered, in the defensive position 1st Battalion, 9th Marines, occupied for a week while waiting for helicopters to extract the guns. From left to right: 2dLt Jerry Jackson (Company C), unidentified Marine in background, 2dLt John Judeson (Company B), 1stLt Wesley Fox (Company A), 2dLt Fritz Werner (Headquarters and Service Company), and 2dLt John Harwood (Headquarters and Service Company).

Company A, 1st Battalion, 9th Marines, Awards for 22 February 1969 Action
MEDAL OF HONOR
First Lieutenant Wesley L. Fox
NAVY CROSS
Second Lieutenant William J. Christman III (posthumous)
First Lieutenant Lee R. Herron (posthumous)
Second Lieutenant George M. Malone
SILVER STAR
Lance Corporal John R. Baird Jr. (posthumous)
Lance Corporal David A. Chacon (posthumous)
Lance Corporal Darrell H. Chapman
Lance Corporal Michael P. Hester
Sergeant David A. Beyerlein
Staff Sergeant Robert R. Jensen
Lance Corporal William C. Northington (posthumous)

Stars and three Navy Crosses awarded. For his actions, Fox received the Medal of Honor.

LAOS INCURSION

While 1st Battalion, 9th Marines, was locked in the biggest contact of Operation Dewey Canyon on 22 February, 3d Battalion, 9th Marines, ground toward Tiger Mountain, where they faced stiffened enemy resistance at the base of the hill from troops in bunkers and spider holes to snipers tied in trees.[227] When 3d Battalion crossed the Da Krong River days earlier, it ran into the People's Army of Vietnam's *812th Regiment* and *18th Battalion*, which moved into the area to check the Marines' advance.[228] Once Companies K and M, 3d Battalion, interdicted Route 548, though, it became clear that the enemy pulled back. Companies I and L, 3d Battalion, took the bomb-damaged Tiger Mountain on 20 February with less trouble than the battalion anticipated.[229] The one-time NVA stronghold was now the site of the last fire support base the Marines constructed during the operation, named Fire Support Base Turnage after General Allen H. Turnage, 3d Marine Division commander in World War II during the Battles of Bougainville and Guam. From its commanding heights, Fire Support Base Turnage sat athwart the Da Krong and A Shau

---

[227] Hyde interview.
[228] LtCol Elliott R. Laine Jr. to Col Robert H. Barrow, "Combat Operations After Action Report," 25 March 1969, attached to 3d Battalion, 9th Marines ComdC, 1 March to 31 March 1969, HD Archive.

[229] 3d Battalion, 9th Marines ComdC, 1 February to 28 February 1969; and Hyde interview.

Defense Department (Marine Corps) A192845

Riflemen from Company L, 3d Battalion, 9th Marines, advance up pockmarked Tiger Mountain, soon to be home of Fire Support Base Turnage.

Valleys and gave Battery E, 2d Battalion, 12th Marines, the ability to support the rifle companies along the Laotian border to the west, where they were cutting Route 922 and disrupting Base Area 611.[230]

The 9th Marines now commanded much of the high ground, but that did not mean they could close with and destroy the enemy. Their lightning advance south during the last 10 days was a master class in how to conduct a regiment-in-the-attack operation, rolling up opposing forces inside the Republic of Vietnam and destroying supply depots and caches. The Laotian border, however, was a political barrier for USMACV, and to conduct combat operations across it could have ramifications. The enemy, however, did not face such obstacles. Laos remained a sanctuary from which North Vietnamese units chose where and when to harass the Marines now moving along Route 922.

Colonel Barrow was dissatisfied that he was unable to protect his right flank, defend against 122mm guns inside Laos, and exploit targets of opportunity.[231] Major General Davis, aware that Barrow was hamstrung, attempted to find ways to support his regimental commander. He had the same fight once before, after the enemy sapper attack on Fire Support

---

[230] 9th Marines ComdC, 1 February to 28 February 1969.

[231] Smith, *High Mobility and Standdown, 1969*, 40–41.

Base Cunningham on 17 February. In the wake of the attack, Davis requested from General Abrams that two battalions from 9th Marines conduct a raid into Laos. USMACV denied the request, citing current rules of engagement. Barrow did learn a valuable piece of information, though. The rules allowed for forces to "maneuver, while actually engaged and in contact with enemy forces, into Laos as necessary for the preservation of the force."[232] If 9th Marines were in contact along the border, they were within their rights to cross into Laos in self-defense. Until those conditions presented themselves, however, Davis had to find other means to protect the regiment's flank. He discussed with XXIV Corps the possibility of USMACV's Studies and Observations Group (USMACV-SOG), a Joint special operations task force, expanding their reconnaissance operations in the Laotian panhandle to include Base Area 611. USMACV approved the request, effective immediately.[233]

It was not until 20 February that companies from 1st and 2d Battalion, 9th Marines, approached the border. While Company A, 1st Battalion, reached the complex at Lang Ha-Bn, Companies E and H from Lieutenant Colonel George Fox's 2d Battalion reached Objective 1 on the regiment's right flank. From a ridgeline on the border, Company H watched an enemy convoy moving west on Route 922, which highlighted a problem Lieutenant Colonel Fox faced: all of Route 922 in 2d Battalion's sector was in Laos, not the Republic of Vietnam. The company commander watching the convoy, Captain David Winecoff, assumed two things about his situation. First, the NVA troops likely knew the Marines were looking down at them from the ridgeline, since his company earlier received a helicopter resupply of 60mm mortar ammunition, which his mortar section now was using to interdict the convoy. Second, the enemy likely believed the Marines would not come across the border, given the political implications. Winecoff reported his observations to battalion and sent out patrols down the ridgeline.[234]

The previous conversations between 3d Marine Division, XXIV Corps, and USMACV now came into play.[235] With the information Winecoff provided, as well as intelligence from the direct support and special operations teams now operating near 9th Marines, Army Lieutenant General Richard G. Stilwell, XXIV Corps commander, contacted USMACV on 20 February, reporting that the enemy was evacuating their heavy artillery from the area of operations. Lieutenant General Stilwell suggested two courses of action: a raid across the border into Base Area 611 not to exceed five kilometers, or a USMACV-SOG operation with one Marine company acting as a quick reaction force. Barrow did not wait for Abram's answer.[236] On the afternoon of 21 February, 9th Marines headquarters ordered 2d Battalion to set up an ambush along Route 922 that night and return across the border by 0630 on 22 February.[237]

Despite the apprehension from higher headquarters about crossing the border, 1st Battalion's Companies A and B had already been upward of 1,000 meters inside Laos. It is unclear if USMACV or even the Marines themselves were aware. Even Winecoff's 2d Platoon, which he had sent down the ridgeline earlier in the day, was also inside Laos, 1,000 meters from Route 922. After receiving his orders, Winecoff spent an hour drawing up his ambush and patrolling order, and he instructed 3d Platoon to remain on the ridgeline with the mortar section. The company command group with 1st Platoon set off at 1630 and joined 2d Platoon an hour later.[238]

Over the next half hour, Winecoff devised a plan for a linear ambush with flank security on both sides and rear security at the rally point. He briefed his squad leaders about the importance of maintaining contact during the night movement and avoiding triggering the ambush prematurely. It was not until Winecoff was already in Laos that Barrow informed higher headquarters what he was doing. During the next hour, he pled his case against much opposition. Finally, Barrow's superiors authorized the ambush. Barrow understood, however, that he might have his approval but "the implication clearly was, 'You better make it work'."[239]

The platoons moved out at 1700, advancing in column over difficult terrain for the next three hours, avoiding trails to maintain secrecy on their approach to Route 922. A reconnaissance of the area took two nerve-wracking hours, as the pitch-black conditions made every fallen log look like an enemy

---

[232] As quoted in Smith, *High Mobility and Standdown, 1969*, 41.
[233] Smith, *High Mobility and Standdown, 1969*, 40–41.
[234] Winecoff interview.
[235] MajGen Davis was in Hong Kong at the time. Davis interview.
[236] Barrow interview.
[237] Smith, *High Mobility and Standdown, 1969*, 41–42.
[238] Winecoff interview.
[239] Barrow interview.

soldier to the Marines. While the rest of the company waited, Winecoff and the two platoons hugged the ground waiting for individual vehicles, including a tracked vehicle with a searchlight, to slowly pass. Once the reconnaissance team considered the ambush site clear, Winecoff ordered his Marines across a stream, up the five-meter bank, and over Route 922.[240]

By 0100, the ambushers were 1,750 meters inside Laos and in position, with 1st Platoon on the left, 2d Platoon on the right, and the command group in the center. Only about 25 meters from the road on a hillside looking north, the two platoons allowed several small convoys to pass during the next two hours, including a solitary soldier on foot doing reconnaissance by fire. On order from Winecoff, his Marines were not to fire their weapons until he detonated his Claymore antipersonnel mine. Winecoff was unsure what they were looking for—all Barrow told him was that there were "good reasons for the ambush," but the regimental commander could not say why.[241] Winecoff was aware that 1st Battalion, 9th Marines, captured two 122mm field guns to his east the day before, but he did not know that the battalion already interdicted Route 922. He deduced that the enemy was attempting to evacuate the other heavy artillery along the route where he set his ambush.

At 0230, engines started to the company's right and drivers turned on black-out lights. Six vehicles, one of them tracked, moved slowly, stopping every 100–150 meters and shutting off engines to do reconnaissance by silence, as had the previous convoys.[242] It took a half hour for the lead vehicle to round a bend in the road, coming into view of the ambushers. The driver shut down the engine to coast down a slope, and Winecoff waited until the convoy was securely in the zone of fire. When the third vehicle entered the zone and stopped, Winecoff, who had his "eyes glued to that spot," detonated his Claymore, killing three enemy soldiers in the second truck. The 1st Platoon on the left opened up on the lead vehicle, which was carrying five tons of small-arms ammunition, followed by the 2d Platoon firing on the third truck. The company's forward observer ordered artillery fire on the road within 30 seconds, providing cover for the company to assault across the road.[243]

With the remnants of the convoy ablaze, the platoons lined up in column and followed more or less their earlier route in reverse, but this time using trails to leave the ambush zone as quickly as possible. After moving 500 meters from Route 922, Winecoff stopped the column in a position that allowed the company to spend the morning of 22 February resting while maintaining observation of the road for potential opportunities to use artillery.[244]

The ambush had limited impact when viewed against the backdrop of interdicting NVA forces' evacuation of Base Area 611. Its importance, however, was in the struggle Barrow and Major General Davis had with USMACV over operating across the border. Later that morning, Barrow reported to higher headquarters that he ordered an ambush into Laos in the hopes of leveraging the success and the value of the mission against USMACV's qualms about similar operations. In his report, Barrow argued that B-52 Arclight missions, fixed-wing strikes, and artillery fires were all unable to prevent the enemy's use of Route 922. This problem increased in importance as 9th Marines approached the border while NVA commanders could still reinforce and resupply their units. In the colonel's mind, this "seriously endangered" the security of his regiment, as did the prospect of the enemy removing the 122mm field guns to positions "from which fires could be delivered on my forces."[245] He closed the report in blunt terms: "Put another way, my forces should not be here if ground interdiction of Route 922 not authorized."[246] Two days later, General Abrams authorized Lieutenant General Stilwell's proposal for operations inside Laos, but with two stipulations: units were not to exceed two kilometers beyond the border, and there would be no public discussion of incursions.[247] Barrow, after three weeks, finally got his authorization to operate across the border with more than platoons. Winecoff's contributions to the success of the mission and his display of leadership resulted in a Silver Star.

---

[240] LtCol David F. Winecoff, "Night Ambush!," *Marine Corps Gazette* 68, no. 1 (January 1984): 49–50.
[241] Winecoff interview.
[242] 2d Battalion, 9th Marines ComdC, 1 February to 28 February 1969.
[243] Winecoff interview; and Winecoff, "Night Ambush!," 51.
[244] Winecoff interview.
[245] Barrow to Davis report.
[246] Barrow to Davis report.
[247] Smith, *High Mobility and Standdown, 1969*, 45.

ROUTE 922 AMBUSH

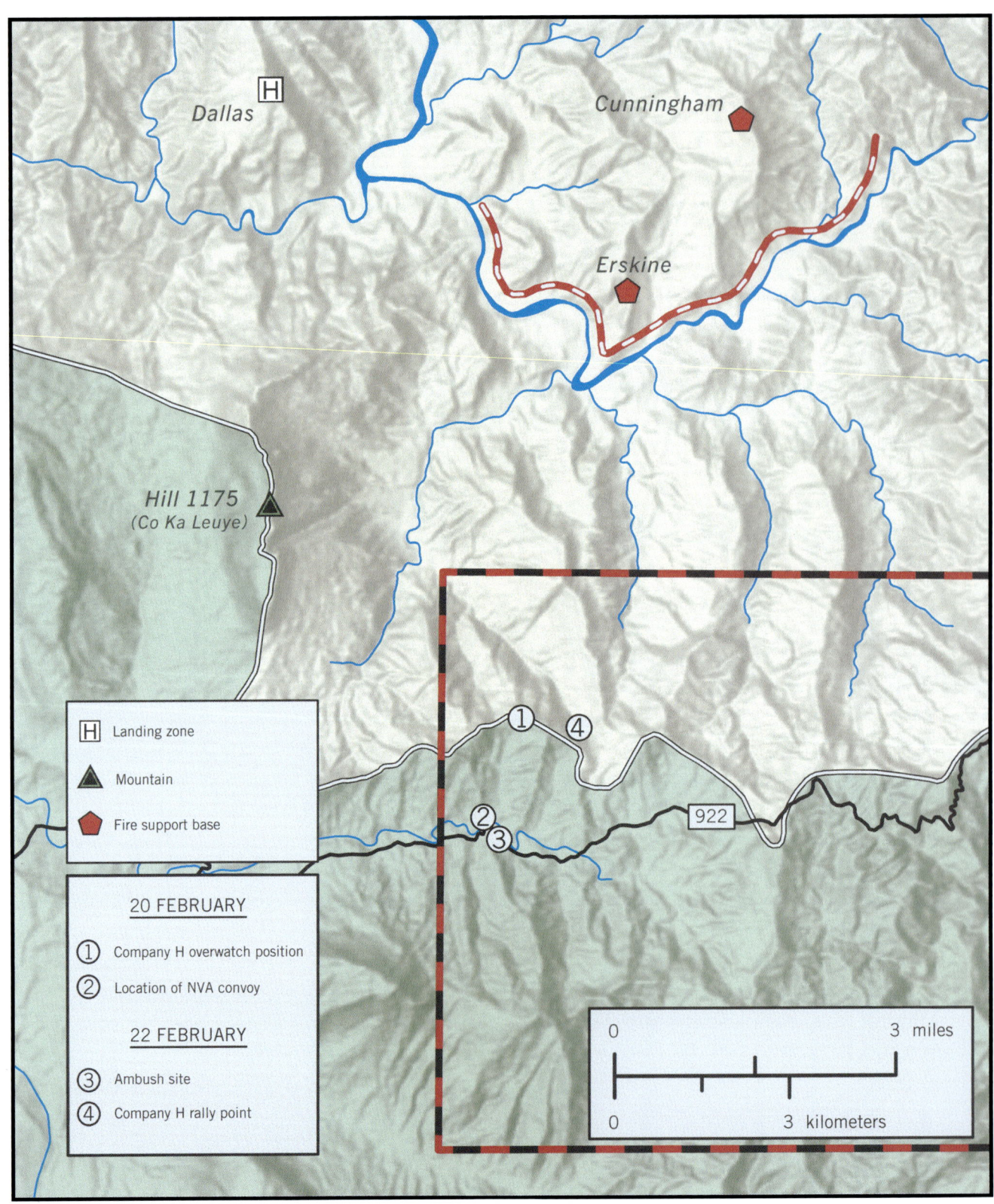

Map courtesy of Pete McPhail, adapted by MCUP

Defense Department (Marine Corps) A192691

Capt Thomas F. Hinkle (left) and Marines from Company M, 3d Battalion, 9th Marines, inspect a spiked 122mm field gun.

HAMMER AND ANVIL

The events on 22 February led to 9th Marines' exploitation of Route 922.[248] Now 1st Battalion had a clear path to two objectives: Hill 1044, which overlooked the road, and Tam Boi, three kilometers to the east and at the intersection of Routes 548 and 922. The gains came at a cost, however. Captain George Meerdink Jr., commander of Company B, 1st Battalion, died along with three in his command group inside the Laotian border from a direct 60mm mortar hit.[249] Several corpsmen attempted to save the wounded in the command group during the barrage of enemy mortars, earning them Bronze Stars.[250]

Only hours after Abrams consented to Barrow's request on 22 February, regimental headquarters ordered Lieutenant Colonel Fox and 2d Battalion to interdict anything that came down Route 922. Captain Winecoff and Company H were enjoying a resupply of food and beer at their original position on the border at 1530 when the command group received orders from battalion to move the entire company back to the ambush site that night.[251] In the interim, Company F, 2d Battalion, arrived overland and took Winecoff's position on the border, where the battalion command post joined them. Helicopters lifted Company G, 2d Battalion, and the 81mm mortar platoon into Laos that afternoon, onto the forward slope of the ridgeline that overlooked Route 922. Fox now positioned his entire battalion inside Laos to serve as the anvil to 1st and 3d Battalions' hammer.[252]

For the next five days, Company H served as the lead element of the 2d Battalion, covering 5,000 meters at a breakneck pace when they were not setting ambushes along Route 922. The operation yielded the results Barrow hoped it would, with Company H intercepting three 122mm field guns, all of which the enemy was attempting to evacuate from the area. The first gun, which the company found on 25 February, was the victim of an air strike; the other two the company discovered on the side of the road. To ensure that the enemy could not salvage the guns, the engineer team from Company C, 3d Engineer Battalion, placed a diamond charge of C-4 explosive on the barrels and breaches as well as several artillery rounds underneath.[253] Aside from the guns, 2d Battalion also found 2.5 million units of penicillin, 6 million units of streptomycin, 20 tons of food, 85mm artillery pieces, 23mm and 12.7mm antiaircraft guns, numerous bunkers, and multiple calibers and amounts of small-arms ammunition. They even found an A-6E Intruder wreckage and the remains of the pilots, likely

---
[248] Smith interview.

[249] 1st Battalion, 9th Marines ComdC, 1 February to 28 February 1969.
[250] 2dLt William W. Chapman, interview with SSgt Willis S. Bernard Jr., 19–28 April 1969, Marine Corps Oral History Collection.
[251] Winecoff interview.
[252] Fragmentary Order 46–69, 22 February 1969, attached to 2d Battalion, 9th Marines ComdC, 1 February to 28 February 1969.
[253] Barrow to Davis report.

TACTICAL SITUATION, PHASE III

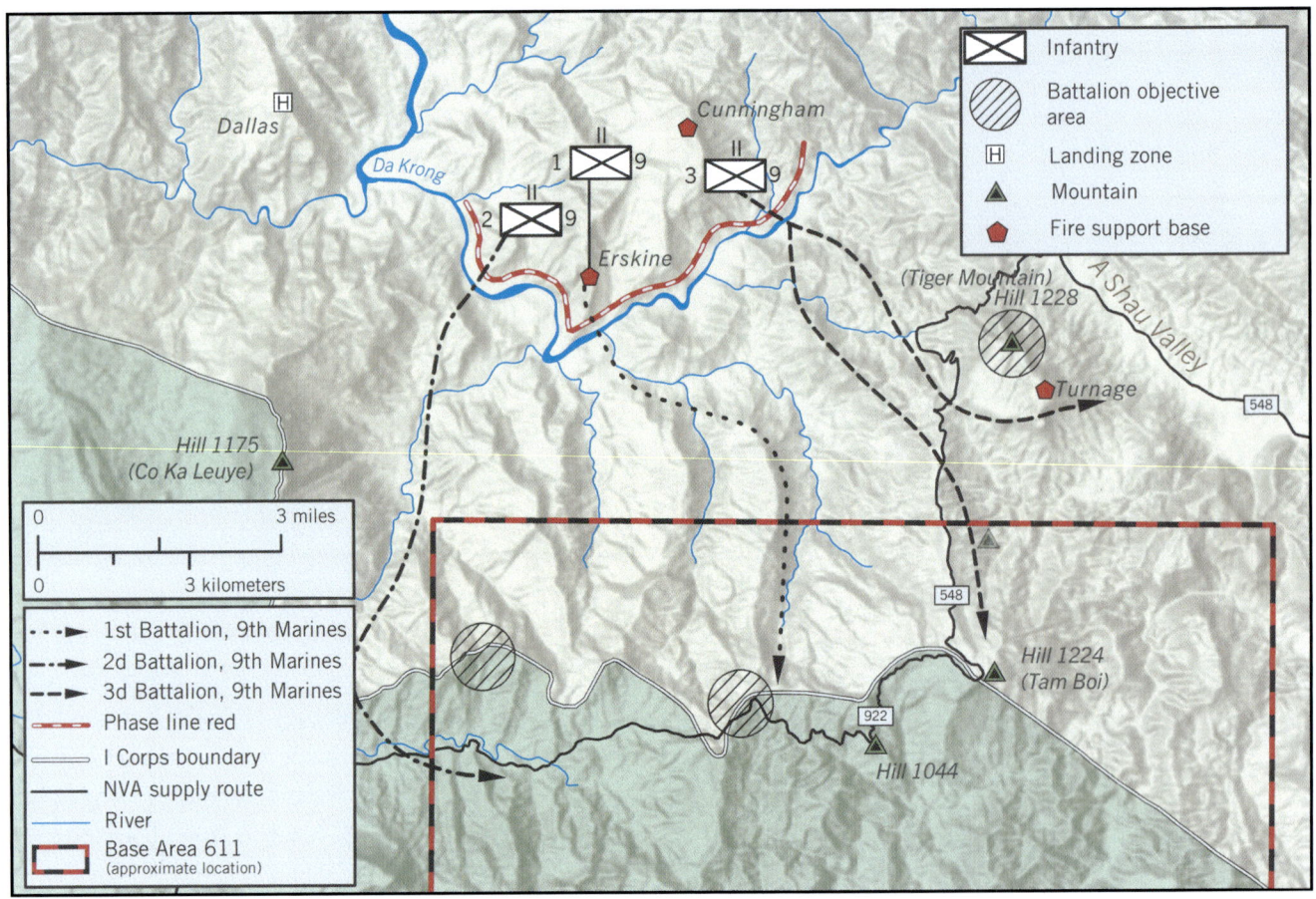

Map courtesy of Pete McPhail, adapted by MCUP

the aircraft that NVA soldiers shot down in December 1968, which alerted 3d Marine Division to increased activity in Base Area 611.[254] The battalion's successes came at the cost of 8 killed and 33 wounded. An enemy ambush on 25 February that claimed three Marines killed and five wounded could have been worse were it not for Corporal William D. Morgan, who singlehandedly assaulted an enemy bunker as a diversion so that the Marines could evacuate their wounded comrades.[255] For his actions, Corporal Morgan posthumously received the Medal of Honor.

With 2d Battalion, 9th Marines' assault down Route 922, the regiment converged near Tam Boi, which served as an enemy headquarters complex and was the refuge of the troops who fled from 3d Battalion, 9th Marines, at Tiger Mountain.[256] Lieutenant Colonel Smith's 1st Battalion, 9th Marines, wheeled left and made a dash east, traveling on Route 922 ahead of 2d Battalion, 9th Marines, with its target Hill 1044, in the shadow of Tam Boi and 500 meters inside Laos. To the north, Companies L and M, 3d Battalion, 9th Marines, marched south together from Tiger Mountain, traveling down Route 548.[257] Along the way, they discovered infrastructure to maintain the important supply route, from garages for construction equipment to bunkers every 15 meters in which engineers could seek shelter during air raids. The companies also discovered they were encountering the People's Army of

---

[254] 2d Battalion, 9th Marines ComdC, 1 February to 28 February 1969. Company H found the identification tags of Navy Lt Gary J. Meyer.
[255] 2d Battalion, 9th Marines ComdC, 1 February to 28 February 1969.

[256] Hyde interview.
[257] Hinkle interview.

Corporal William D. Morgan

Medal of Honor Citation

The President of the United States of America, in the name of Congress, takes pride in presenting the Medal of Honor (Posthumously) to Corporal William David Morgan (MCSN: 2337025), United States Marine Corps, for conspicuous gallantry and intrepidity at the risk of his life above and beyond the call of duty while serving as a squad leader with Company H, in operations against the enemy. While participating in Operation DEWEY CANYON southeast of Vandegrift Combat Base, one of the squads of Corporal Morgan's platoon was temporarily pinned down and sustained several casualties while attacking a North Vietnamese Army force occupying a heavily fortified bunker complex. Observing that 2 of the wounded Marines had fallen in a position dangerously exposed to the enemy fire and that all attempts to evacuate them were halted by a heavy volume of automatic weapons fire and rocket-propelled grenades, Corporal Morgan unhesitatingly maneuvered through the dense jungle undergrowth to a road that passed in front of a hostile emplacement which was the principal source of enemy fire. Fully aware of the possible consequences of his valiant action, but thinking only of the welfare of his injured companions, Corporal Morgan shouted words of encouragement to them as he initiated an aggressive assault against the hostile bunker. While charging across the open road, he was clearly visible to the hostile soldiers who turned their fire in his direction and mortally wounded him, but his diversionary tactic enabled the remainder of his squad to retrieve their casualties and overrun the North Vietnamese Army position. His heroic and determined actions saved the lives of 2 fellow Marines and were instrumental in the subsequent defeat of the enemy. Corporal Morgan's indomitable courage, inspiring initiative and selfless devotion to duty upheld the highest traditions of the Marine Corps and of the U.S. Naval Services. He gallantly gave his life for his country.

Defense Department
(Marine Corps) A700693

Vietnam's *559th Transportation Group*.[258] Simultaneous to 9th Marines' battalions converging on the southeastern corner of the area of operations, elements of the 1st Brigade, 101st Airborne Division, were farther to the east on the other side of the A Shau Valley conducting a sweep of Base Area 114 in Operation Spokane Rapids from 20 February to 3 March.[259] With the combination of the two operations on either side of the A Shau Valley, III MAF forces effectively cut off NVA units in the Republic of Vietnam from their base areas.

On 24 February, First Lieutenant Benfatti's Company L, 3d Battalion, 9th Marines, and Captain Thomas Hinkle's Company M, 3d Battalion, 9th Marines, took Tam Boi. As with Tiger Mountain, the enemy preferred to pull back and probe at night than defend their installations in a pitched battle. Barrow was aware of enemy discussions about Tam Boi's vulnerability because of the direct support team's intercepts. Confirming Tam Boi's importance, Barrow requested a B-52 strike on the area on 20 February.[260] When the two companies assaulted the hill four days later, they found two spiked 122mm field guns oriented toward the A Shau Val-

---

[258] Hyde interview.
[259] David Burns Sigler, *Vietnam Battle Chronology: U.S. Army and Marine Corps Combat Operations, 1965–1973* (Jefferson, NC: McFarland, 1992), 91.

[260] Rayburn, "Direct Support during Operation DEWEY CANYON," 18.

Jonathan F. Abel Collection, Archives Branch, Marine Corps History Division

PFC Bernardo A. Blazek stands atop a destroyed Soviet-made heavy artillery prime mover.

ley, three prime movers, large ammunition caches, and training areas. A comprehensive tunnel and cave complex surprised the Marines, who found tunnels between 40–250 meters long dug into solid rock that could survive direct hits from air and artillery attacks.[261]

Company D, 1st Battalion, 9th Marines, discovered on 27 February that the enemy could take advantage of the devastation that Arclight missions wrought. When taking Hill 1044, they uncovered the largest arms and ammunition cache of the war to date. Company commander Captain Edward F. Riley had some indication he was facing a weapons storage area when assaulting Hill 1044 on 26 February, as air strikes set off several large, secondary explosions. The next morning, Captain Riley sent out a platoon-size patrol from the crest of the hill that immediately unearthed 18 machine guns.[262] Throughout the day, the scale of the supply depot became apparent. Along with prepared bunkers, enemy troops used the gigantic bomb craters from Arclight missions as makeshift storage areas, covering more than 100 tons of munitions and hundreds of small arms with dirt. The job was so large that Lieutenant Colonel Smith ordered the rest of 1st Battalion to assist. Company C arrived several days later, sweeping down Route 922 on the way, and immediately discovered a 50-ton ammunition dump in a narrow saddle feature of Hill 1044, with easy access to the main supply route.[263] When First Lieutenant Wesley

---

[261] Barrow to Davis report; and Hyde interview.

[262] Riley interview.
[263] Kelly interview.

ENEMY BUNKER AND WEAPONS CACHE LOCATIONS

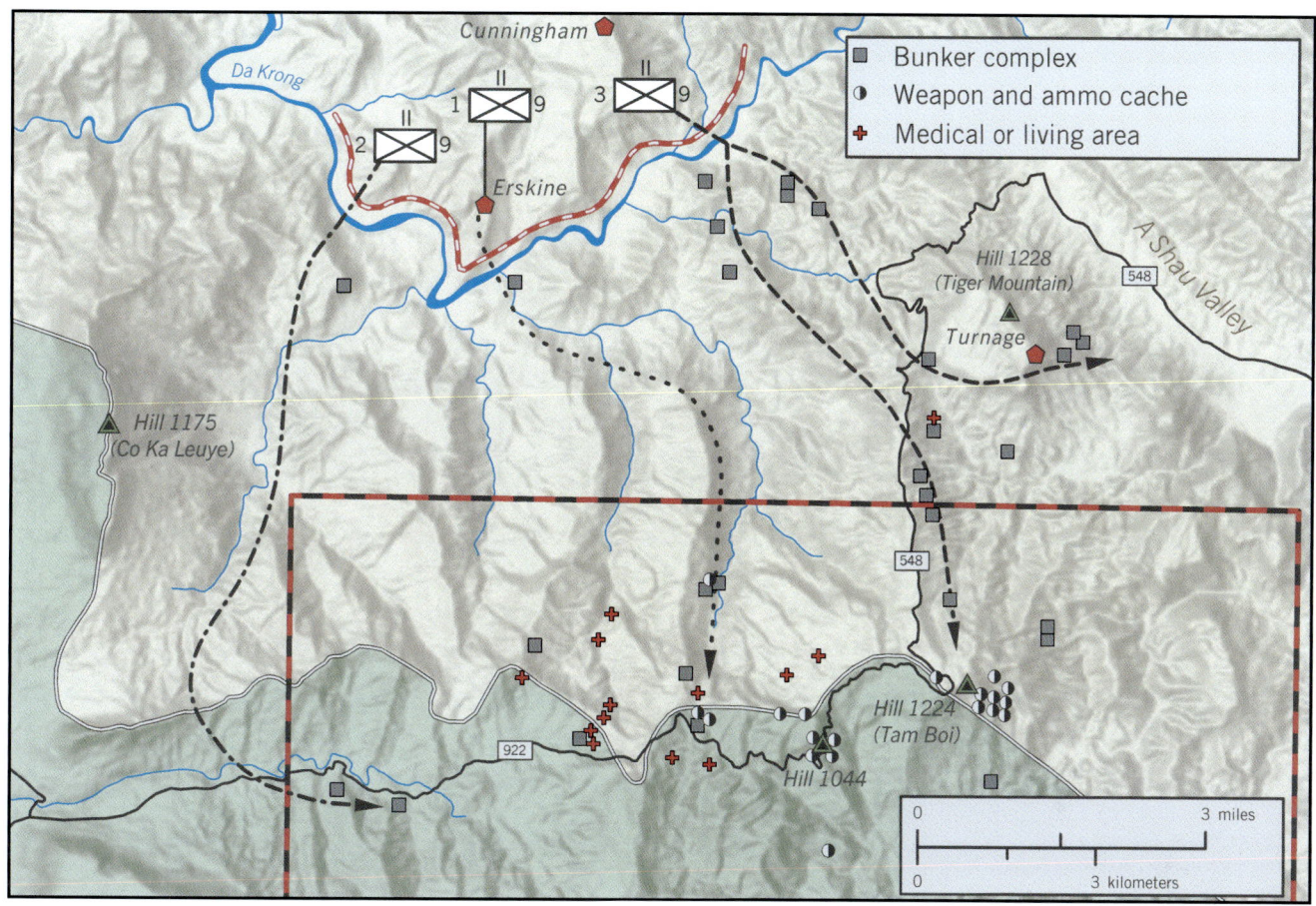

Map courtesy of Pete McPhail, adapted by MCUP

Fox arrived with Company A, the size of the cache, which was equivalent to what a Marine force service regiment would hold, staggered him. He reported to Smith that the only way to measure it was by tons. Smith believed such a metric was unsatisfactory and ordered his Marines to count each piece, instruction that Fox thought was both "hopeless" and "useless." The official tally after two days—629 rifles, 60 machine guns, 19 antiaircraft guns, etc.—Fox admitted was an undercount.[264]

EXTRACTION

For the next two weeks, the battalions conducted sweeps, many of which were inside Laos, and uncovered more caches that the engineers from Company C, 3d Engineer Battalion, and the explosive ordnance disposal platoon from Force Logistic Support Group Bravo destroyed.[265] Meanwhile, elements of 2d Brigade, 101st Airborne Division, conducted a reconnaissance in force operation into the A Shau Valley to interdict Route 548 called Operation Massachusetts Striker, which lasted into early May.[266]

The enemy's 1969 Tet Offensive began on 23 February, the day USMACV predicted. There were marked differences between it and its predecessor. The offensive was more intense, totaling one-third more attacks than 1968, but it focused only on the I and III Corps Tactical Zones,

[264] Fox interview.
[265] Barrow to Davis report.
[266] Shelby L. Stanton, *The Rise and Fall of an American Army: U.S. Ground Forces in Vietnam, 1965–1973* (New York: Random House, 1995), 295–96.

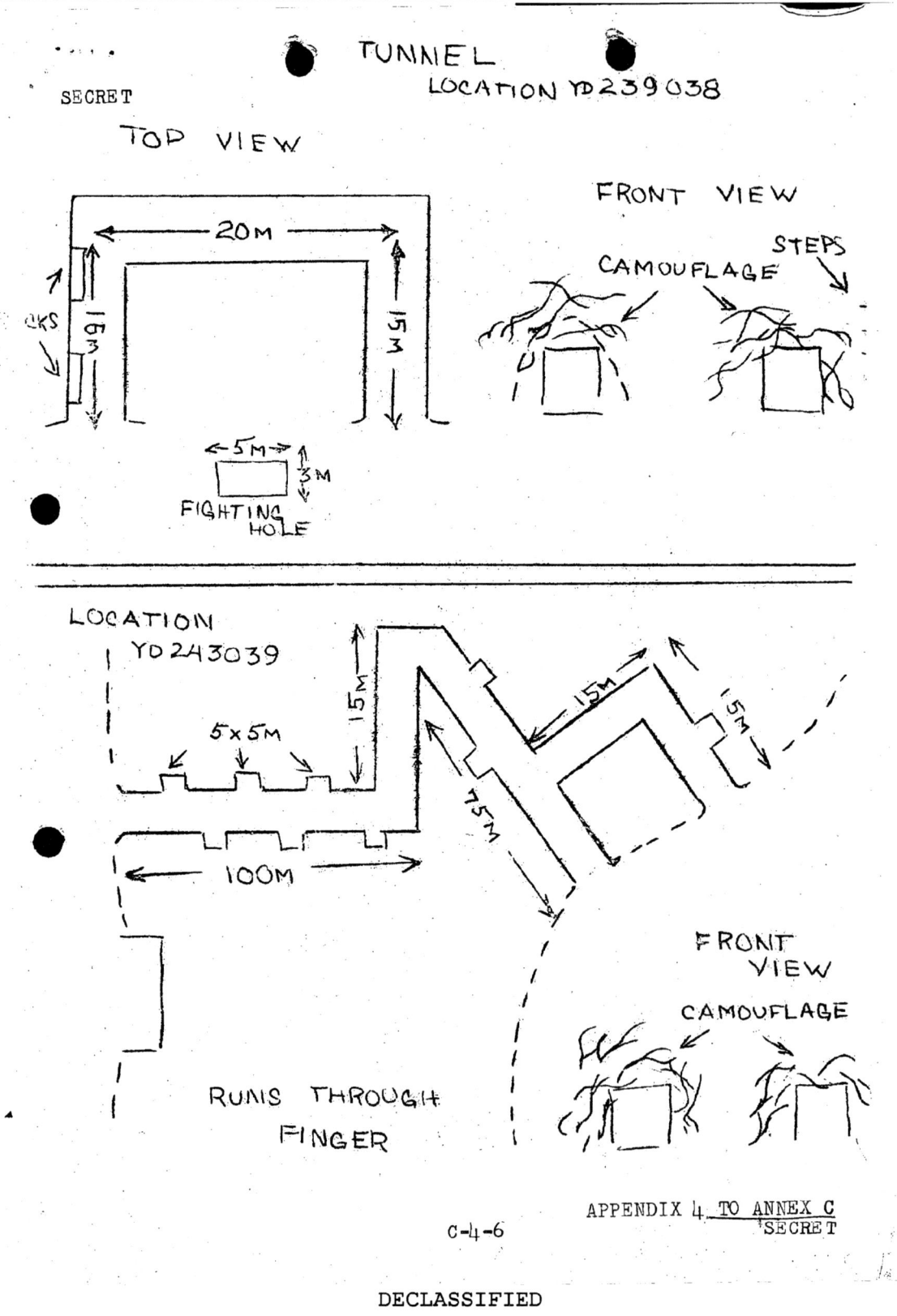

Official 9th Marines' sketches of two tunnel complexes at Tam Boi.

Marines from Company D, 1st Battalion, 9th Marines, assemble a captured enemy 12.7mm antiaircraft gun, one of the many weapons they uncovered in the reported 350-ton cache on Hill 1044.

not a nationwide general uprising. These attacks were almost entirely NVA troops, not local forces, and were against military facilities and units near Da Nang and Saigon, not population centers. In I Corps, rocket and mortar attacks targeted command and logistic facilities at Da Nang and Chu Lai, and Marines repelled predawn ground attacks against their compounds.[267]

It is difficult to determine if Operation Dewey Canyon blunted the 1969 Tet Offensive given that the enemy's objectives appeared limited relative to Tet 1968. What is certain, however, is that American forces were better prepared in 1969, owing not only to better intelligence but operations designed to keep pressure on the enemy, highlighting once again the connection between conventional and pacification efforts. General Abrams stressed this point in preparation for the winter-spring offensive. To him, it was crucial to "beat the hell out of the cadre and local forces" while attacking main

---

[267] Smith, *High Mobility and Standdown, 1969*, 97–100.

forces as well as base areas and supply points. Doing both would make it militarily impractical for the enemy to operate. "This," Abrams contended, "is the real meat and potatoes."[268] Operation Dewey Canyon did that, as did Operation Taylor Common near Hoi An in Quang Nam Province. Using high mobility along the coastline, 1st Marine Division's Task Force Yankee attacked Base Area 112, a mountain sanctuary for NVA units that Marines dubbed the Arizona Territory. Beginning 7 December 1968, Operation Taylor Common occurred nearly simultaneous to Operation Dewey Canyon, and by 8 March 1969 achieved the objective of neutralizing Base Area 112 to protect Da Nang while destroying an NVA regiment and capturing tons of food, weapons, and ammunition.[269]

At the same time, the conclusion of Operation Dewey Canyon came into view for 9th Marines and 2d Battalion, 12th Marines. The Marines severed Route 922 and destroyed a majority of Base Area 611's depots on or near the border. By all measures, 9th Marines achieved its mission of disrupting enemy operations in the Da Krong and A Shau Valleys and the enemy's ability to supply forces deeper inside the Republic of Vietnam. Among their accomplishments, the regiment seized or destroyed two medium artillery batteries, one light artillery battery, the enemy's antiaircraft capabilities in the region, and headquarters and storage facilities of an estimated regiment.[270] In the first week of March, they prepared a phased retraction.

Barrow and his planners envisioned the withdrawal as a reverse of the leapfrogging actions in Operation Dewey Canyon's three phases: rifle companies would return to the fire support bases under the protective cover of artillery, and helicopters would lift the elements to Vandegrift Combat Base.[271] All told, the retrograde operation was to take seven days, with 2d Battalion, 9th Marines, lifting out on 3 March, 1st Battalion, 9th Marines, on 4 March, and the ARVN battery and final company from 3d Battalion, 9th Marines, on Fire Support Base Turnage leaving on 7 March. Helicopters lifted 2d Battalion, 9th Marines, minus Company G to Vandegrift Combat

Barry Broman Collection, Archives Branch, Marine Corps History Division

A CH-46 extracts Company H, 2d Battalion, 5th Marines, near the end of Operation Taylor Common, February 1969.

Base on schedule before three issues forced a change in plans: inclement weather, a lack of artillery ammunition to protect the withdrawal, and the need for 9th Marines to relieve 3d Marines in the field.[272]

Due to the changes, regimental headquarters tasked 1st Battalion, 9th Marines, with extracting from Laos the USMACV-

---
[268] Quoted in United Press International, "US Commander in Vietnam Wants No US Troops Sent Home Until Success of War Is More Certain," *Sacramento (CA) Bee*, 13 January 1969.
[269] Smith, *High Mobility and Standdown, 1969*, 80–94.
[270] Smith, *High Mobility and Standdown, 1969*, 48–49.
[271] Barrow interview.

---
[272] Davis to Cushman, "Artillery Report of Operation Dewey Canyon," 5 May 1969, box 9, folder 19, Vietnam War Collection, HD Archive.

Private First Class Alfred M. Wilson
Medal of Honor Citation

The President of the United States of America, in the name of Congress, takes pride in presenting the Medal of Honor (Posthumously) to Private First Class Alfred Mac Wilson (MCSN: 2421744), United States Marine Corps, for conspicuous gallantry and intrepidity at the risk of his life above and beyond the call of duty while serving as a Rifleman with Company M, Third Battalion, Ninth Marines, Third Marine Division in action against hostile forces in the Republic of Vietnam. On March 3, 1969, while returning from a reconnaissance in force mission in the vicinity of Fire Support Base Cunningham in Quang Tri Province, the First Platoon of Company M came under intense automatic weapons fire and a grenade attack from a well-concealed North Vietnamese Army force pinning down the center of the column. Rapidly assessing the situation, Private First Class Wilson, acting as Squad Leader, skillfully maneuvered his squad to form a base of fire and act as a blocking force while the point squad moved to outflank the enemy. During the ensuing fire fight, both his machine gunner and assistant machine gunner were seriously wounded and unable to operate their weapon. Realizing the importance of recovering the M-60 machine gun and maintaining a heavy volume of fire against the hostile force, Private First Class Wilson, with complete disregard for his own safety, followed by another Marine, fearlessly dashed across the fire-swept terrain to recover the weapon. As they reached the machine gun, a North Vietnamese soldier threw a grenade at the Marine. Reacting instantly, Private First Class Wilson fired a burst from his M-16 rifle killing the enemy soldier. Observing the grenade fall between himself and the other Marine, Private First Class Wilson, fully realizing the inevitable result of his actions, shouted to his companion and unhesitatingly threw himself on the grenade, absorbing the full force of the explosion with his own body. His heroic actions inspired his platoon members to maximum effort as they aggressively attacked and defeated the enemy. Private First Class Wilson's indomitable courage, inspiring valor and selfless devotion to duty upheld the highest traditions of the Marine Corps and the United States Naval Service. He gallantly gave his life for his country.

SOG forces that assisted in the 22 February incursion.[273] The battalion also spent several extra days destroying the caches on Hill 1044 before moving to Tam Boi and Fire Support Base Turnage throughout the second week of March, where it swept the immediate area and discovered yet more caches. In the meantime, 3d Battalion, 9th Marines, pulled back to Fire Support Base Cunningham to wait for the bad weather to break, again relying on C-130s for resupply drops.[274] It was in this period of retrograde actions that Private First Class Alfred M. Wilson from Company M, 3d Battalion, 9th Marines, smothered a grenade during an enemy ambush to protect another Marine. For his actions, Wilson received a posthumous Medal of Honor, the final of Operation Dewey Canyon.

Conditions became reminiscent of the previous month, where poor weather grounded casualty evacuation and resupply helicopters. This kept the companies from ranging too far from the relative safety of the fire support bases but in a constant state of patrolling and ambushes to keep the enemy at arm's length. As delays in extraction continued, which Captain John Kelly described as "a matter of long waiting," the enemy reversed tactics.[275] For much of Operation Dewey Canyon, the Marines pushed the North Vietnamese main force ahead of them. With the extraction underway, the enemy returned to the Republic of Vietnam from Laos and ringed the fire support bases, maintaining pressure on the Marines.[276] Throughout, enemy troops employed what seemed to 1st Battalion, 9th Marines, operations officer Major James P. McWilliams

---

[273] Smith, *High Mobility and Standdown, 1969*, 49.
[274] 1stLt John W. Roth, interview with SSgt Willis S. Bernard Jr., 18 March 1969, Marine Corps Oral History Collection.
[275] Kelly interview.
[276] Kelly interview.

First Lieutenant James K. Murphy
Silver Star Citation

The President of the United States of America takes pleasure in presenting the Silver Star to First Lieutenant James K. Murphy (MCSN: 0-104702), United States Marine Corps, for conspicuous gallantry and intrepidity in action while serving as a Platoon Commander with Company L, Third Battalion, Ninth Marines, THIRD Marine Division (Rein.), FMF, in connection with combat operations against the enemy in the Republic of Vietnam. On 18 March 1969, First Lieutenant Murphy's platoon was providing security from commanding terrain overlooking a landing zone at which the Third Battalion was awaiting extraction to the Vandegrift Combat Base when the unit came under a vicious ground and mortar attack from a numerically superior North Vietnamese Army force. Aware of the importance of denying the vital terrain to the enemy, First Lieutenant Murphy completely disregarded his own safety as he moved from one squad to another to ascertain the welfare of his men, pinpoint targets for their suppressive fire and ensure that all possible avenues of hostile approach were adequately covered by firepower. Although stunned by the tenaciousness of the Marine defense, the enemy continued its attack against the platoon and additionally commenced delivering mortar fire on the landing zone as well as automatic weapons fire at approaching extraction helicopters. Realizing the need for immediate action to prevent loss of Marine lives and damage to the aircraft, First Lieutenant Murphy unhesitatingly moved to the forward slope of the hill and, heedless of the hostile rounds impacting all around him, skillfully adjusted supporting artillery fire on North Vietnamese Army mortar emplacements, effectively silencing most of the hostile fire. After the battalion was extracted, the platoon started to deploy to the landing zone for its embarkation but the remaining enemy mortars increased the intensity of their fire and First Lieutenant Murphy, again valiantly moving to a dangerously exposed vantage point, adjusted the bombing and strafing runs of friendly aircraft now on station which destroyed the sources of hostile fire and enabled his men to be extracted in safety. His heroic and determined actions inspired all who observed him and were instrumental in permitting the extraction mission to be accomplished with a minimum of Marine casualties. By his courage, superb leadership, and unwavering devotion to duty in the face of grave personal danger, First Lieutenant Murphy upheld the highest traditions of the Marine Corps and of the United States Naval Service.

---

an "almost inexhaustible supply of 60mm and 82mm mortar ammunition" that they drew from caches the regiment did not find.[277]

The direct support team played one final role in Operation Dewey Canyon during the withdrawal. It intercepted enemy radio traffic about plans to attack the Marines while they loaded helicopters, knowing this would be when they were most vulnerable. With this intelligence, Barrow ordered continuous air support and artillery fire to protect the withdrawal, both of which were crucial in repelling enemy attacks at Tam Boi and Fire Support Bases Cunningham and Turnage. At Fire Support Base Cunningham, 3d Battalion, 9th Marines, and the regimental command post contended with one remaining 122mm field gun emplaced on the heights of Co Ka Leuye, which made the Marines get used to "doing a lot of jumping in holes and bunkers," Second Lieutenant Dallas M. Hyde remembered.[278] Knowing that the enemy bracketed Fire Support Base Cunningham's landing zone, Barrow ordered a new area cleared 300 meters north for the extraction. On 16 March, he received an intercept that the 122mm crew readjusted to the new landing zone for the extraction that day, which several bracketing rounds confirmed. Barrow therefore allowed the enemy to think that the Marines would use the second landing zone. When the helicopters approached for the final evacuation of Fire Support Base Cunningham, they

---

[277] Maj James P. McWilliams, interview with SSgt Willis S. Bernard Jr., 19–28 April 1969, Marine Corps Oral History Collection.

[278] Hyde interview.

landed at the primary landing zone. By the time the 122mm crew readjusted, 3d Battalion, 9th Marines, and the regimental command group were loaded and gone.[279]

Next to leave was 1st Battalion, 9th Marines, and Battery W, 2d Battalion, 12th Marines, on Tam Boi.[280] Barrow knew about a 75mm recoilless rifle and mortars that the enemy planned to use to block the landing zone on Tam Boi when the first helicopters arrived on 18 March. When the lift started at 1230, ordnance rained down on the enemy positions. Simultaneously, enemy mortars opened up on the landing zone. The command post of 3d Platoon, Company C, 1st Battalion, 9th Marines, took a direct hit, seriously wounding the platoon commander and sergeant as well as a radio operator and corpsman.[281] Around 1400, Army CH-47s arrived to assist the Marine CH-46s, quickening the pace of the withdrawal. Army and Marine pilots and crews exhibited skill and courage in the barrage, continuing to land and wait until fully loaded before leaving while mortar rounds fell only meters away. During the extraction, a platoon from Company C, 1st Battalion, 9th Marines, occupied the heights on Tam Boi, using it to spot enemy positions and deny NVA troops the high ground overlooking the landing zone. The platoon beat back an enemy attempt to take the hill and called in air and artillery strikes on eight mortar and four machine-gun positions.[282] By 1630, the entire 1st Battalion, 9th Marines, retrograded without the loss of a Marine or helicopter.[283]

The final extraction was Company I, 3d Battalion, 9th Marines, and Battery E, 2d Battalion, 12th Marines, on the afternoon of 18 March, both of which were at Fire Support Base Turnage to protect the withdrawal from Tam Boi. Again, fixed-wing aircraft played a crucial role in suppressing enemy mortar attacks. Company I provided security while Battery E lifted out their howitzers and loaded onto helicopters. At 1820, the riflemen left Tam Boi, terminating Operation Dewey Canyon.[284]

Defense Department (Marine Corps) 3D1604669

At Fire Support Base Cunningham, Col Robert H. Barrow pins first lieutenant bars on Gordon M. Davis, executive officer of Company K, 3d Battalion, 9th Marines, while MajGen Raymond Davis, 1stLt Davis's father and commanding general of 3d Marine Division, looks on.

CONCLUSION

On 14 April 1969, at Dong Ha, Major General Davis handed command of 3d Marine Division to Major General William K. Jones. The new commanding general maintained Davis's emphasis on seizing enemy caches and blocking NVA units' infiltration into the Republic of Vietnam. High mobility remained the division's method of executing those tasks, but there would be neither an offensive to the same scale as Operation Dewey Canyon, nor would one enjoy the same success. Operation Dewey Canyon is the best-known example of high mobility, but that is perhaps due to its statistical success, which raises its visibility over the innovative operations that preceded it.[285] In fact, the latter phases demon-

---

[279] Rayburn, "Direct Support during Operation DEWEY CANYON," 19–20. See also Barrow interview.
[280] 9th Marines ComdC, 1 March to 31 March 1969, HD Archive.
[281] Kelly interview.
[282] Kelly interview.
[283] Smith interview.
[284] 9th Marines ComdC, 1 March 1969 to 31 March 1969.

[285] Ross Phillips, "Operation Dewey Canyon: Search and Destroy in the Age of Abrams" (master's thesis, Texas A&M University, 2019).

The 9th Marines

Presidential Unit Citation

The assigned and attached units of the 9th Marine Regiment, 3d Marine Division distinguished themselves by extraordinary heroism, professionalism, and achievement in military action against the North Vietnamese Army in the Da Krong and Northern A Shau Valleys, Quang Tri Province, Republic of Vietnam, during the period 22 January to 18 March 1969. Launched under the code name "Dewey Canyon," the concept of the eight-week offensive was a thrust deep into the enemy's rear area to destroy a heretofore impregnable major base. A week of inclement weather delayed the surprise attack and halted combat support, resupply, and medical evacuation. When the weather cleared to the extent that helicopter operations were possible, the Marines found that the enemy had skillfully exploited the delay to strengthen his defense, position medium range artillery, and otherwise complete preparation of the battlefield. Undaunted, the Marines and their South Vietnamese counterparts drove southward, enduring intense artillery, mortar, and automatic weapons fire, to rout a determined enemy from his fortifications. The Marines repulsed numerous counterattacks in the process of decimating the equivalent of two North Vietnamese Army regiments. The 9th Marine Regiment and its attached units destroyed the enemy's regional command control apparatus, eliminated a series of headquarters establishments, and inflicted over 1600 casualties on the North Vietnamese Army. The enemy's engineer and transport capabilities were severely diminished. Additionally, the Marines captured over 1000 tons of weapons, equipment, and supplies; including individual weapons, Infantry crew-served weapons, antiaircraft guns, Field Artillery pieces, vehicles, small-caliber ammunition, and rice. As a result of their gallant actions, the North Vietnamese Army Spring Offensive in the I Corps Tactical Zone was preempted. This magnificent feat of arms, achieved against severe odds and seemingly insurmountable obstacles was made possible by the extraordinary courage, skill, cohesion, and fighting spirit of the 9th Marine Regiment, 3d Marine Division and its attached units. The superb performance of the officers and men of this force represents the essence of professionalism, is in keeping with the highest traditions of the military service, and reflects great credit on them and the Armed Forces of the United States.

---

strated classic fire and maneuver of a regiment in the attack. There were, however, innovative aspects of the operation to go with its foundation in high mobility, from close coordination between the air-ground team to new concepts in logistical support.

Operation Dewey Canyon's success, Barrow later analyzed, was rooted in several factors. First was Davis's notion of carrying the attack to the enemy, wherever and whenever they appeared. Attacking the enemy without observing the principles of war—the objective, the offense, mass, economy of force, surprise, and mobility—would not have guaranteed success, however. Barrow also learned that unit integrity was crucial. After July 1968, 9th Marines did not assign its battalions to other units. This gave time for cohesion and spirit to grow. In that time, the regiment operated in the rugged terrain and dense jungles of Quang Tri Province, which conditioned the Marines to the necessary mental and physical rigors as well as the techniques and skills of mountain warfare.[286] During Operation Dewey Canyon, battalion commanders remained with their companies at all times, walking everywhere their Marines walked and sleeping wherever they slept. Barrow found that this had the effect of making the command post as mobile as the battalion, increasing mutual understanding within the unit, and allowing for quick reactions to tactical situations.[287]

For these reasons, it could be concluded that Marine losses in Operation Dewey Canyon otherwise might have extended

---

[286] Barrow interview. See also Barrow to Davis report.
[287] Command and Staff College, "Notes from Previous Dewey Canyon Symposiums."

MajGen William K. Jones (left) assumes command of 3d Marine Division from MajGen Raymond Davis, 14 April 1969.

beyond the 130 Marines killed and 920 wounded by the conclusion of the operation.[288] It was the professionalism of the Marines involved in the operation that impressed XXIV Corps commander Lieutenant General Stilwell the most. When he wrote his report to General Abrams, he concluded that Operation Dewey Canyon ranked "with the most significant undertakings of the Vietnam conflict." The 9th Marines conducted an operation where no other major American or Republic of Vietnam force operated. Marine helicopter squadrons and U.S. Army helicopter companies sustained a regiment in the field whose base camp was more than 50 linear kilometers away, and an artillery battalion provided almost constant supporting fires. The actions of the Marines and corpsmen in the mountains of Quang Tri Province between 22 January and 18 March 1969 also bear out the significance of the operation. Though an imperfect metric, awards tend to depict the scale of danger, bravery, skill, and sacrifice in an operation. In Operation Dewey Canyon, Marines and corpsmen received 4 Medals of Honor, 6 Navy Crosses, and 55 Silver Stars. For its outstanding performance, 9th Marines earned a Presidential Unit Citation. In Stilwell's estimation, what made Operation Dewey Canyon a resounding success could be ascertained from 9th Marines' "extraordinary cohesion" and "skill in mountain warfare." Just as important, however, was "plain heart."[289]

---

[288] LtGen Stilwell to Gen Abrams, as quoted in Smith, *High Mobility and Standdown, 1969*, 51.

[289] LtGen Stilwell to Gen Abrams, as quoted in Smith, *High Mobility and Standdown, 1969*, 51.

Acronyms and Abbreviations

AFN	American Forces Network
ARVN	Army of the Republic of Vietnam
CHECO	Contemporary Examination of Historical Operations
ComC	Command Chronology
DIA	Defense Intelligence Agency
DMZ	Demilitarized zone
FMF	Fleet Marine Force
HD	History Division
HML	Marine Light Helicopter Squadron
HMM	Marine Medium Helicopter Squadron
I Corps	I Corps Tactical Zone
MAF	Marine Amphibious Force
MAW	Marine Aircraft Wing
MEB	Marine Expeditionary Brigade
MCSN	Marine Corps Service Number
NARA	National Archives and Records Administration
NMMC	National Museum of the Marine Corps
NSN	Navy Service Number
NVA	North Vietnamese Army
Rein	Reinforced
ROTC	Reserve Officers Training Course
TTU	Texas Tech University
VMA	Marine Attack Squadron
VMO	Marine Observation Squadron
USMACV	U.S. Military Assistance Command, Vietnam
USMACV-SOG	U.S. Military Assistance Command, Vietnam–Studies and Observations Group
USMC	United States Marine Corps
USMCR	United States Marine Corps Reserve

Acknowledgments

This commemorative volume is a product of assistance from colleagues within the Marine Corps history program. At the Marine Corps History Division, Dr. Fred Allison and Yvette House provided valuable oral histories. Alisa Whitley and Alyson Mazzone located documents and Samantha Mayo supplied photographs. Dr. Ed Nevgloski shared his knowledge of the topic. Paul Westermeyer and Dr. Breanne Robertson reviewed drafts. Kater Miller and Bruce Allen at the National Museum of the Marine Corps contributed documents and photographs. At Marine Corps University Press, Angela Anderson, Christopher Blaker, and Jason Gosnell provided their editing knowledge and Robert Kocher his design skill. Pete McPhail, with assistance from Grace Stephan and Bret Rodgers, offered his expertise and designed the maps. Ross Phillips reviewed the manuscript. All mistakes are the author's alone. Special recognition goes to the veterans of Operation Dewey Canyon.

Seth A. Givens, PhD
Marine Corps History Division

Seth A. Givens
Marine Corps History Division

Photo courtesy of Stephen Collins,
Leatherneck, adapted by MCUP

Seth A. Givens is a historian at the Marine Corps History Division. He earned a PhD in military history from Ohio University in Athens. At the History Division, his focus is Marines in the Vietnam War and Operation Iraqi Freedom.

Positive Affirmation Posters

A Course in Miracles 1

Words To Help You Change Your Reality

John Vincent Palozzi

Positive Affirmation Posters: A Course in Miracles 1
Words To Help You Change Your Reality

Copyright © 2020 by John Vincent Palozzi

All Rights Reserved

ISBN: 9798649654388

Published by
Palozzi Products
PO Box 1434
Lake Worth FL 33460

www.johnvincentpalozzi.com

johnvincentpalozzi@gmail.com

Foreword

This book contains ten Affirmation Posters meant to be used for inspirational purposes. It is the first "A Course in Miracles" edition and the fifth in the series.

Each poster page is printed on one side.

By opening the book to the page you wish to remove, just push down the page with your hands while the book is flat on a table and "crack" the spine. Turn the page to its back side and do the same. There is a thin layer of glue holding the pages together in the spine of the cover, and by "cracking" it the page will become loose enough to pull it from the book. Do it gently, so as not to rip the paper.

Each page you pull out should measure 8 X 10 inches.

You can now tape it or pin it to the wall, mirror, refrigerator, cupboard door, or wherever you will see and be reminded of the message it gives every time you pass it by. By looking at it often and repeating the message you are reinforcing the thought, which, if you clear your mind and focus on it, along with feeling it, you should find the positive manifestations showing up in your life.

The posters were made to be 8 X 10 on purpose, as that is a very common and available frame size. Whether an inexpensive frame from a dollar store or yard sale, or a more ornate and expensive one, you can frame your poster behind glass or plastic to hang.

You can frame up all ten of the posters and put them around the house, or decorate a whole wall with them. They also can be given as gifts.

The sayings on these posters come directly from the text of A Course In Miracles. Hopefully they will inspire you to investigate The Course further.

Search for other volumes of Affirmation Posters under my name.

Let there be Light,
John

Explanation

This is Volume One of four devoted to a book called "A Course In Miracles." The Course (ACIM) originated in the 1970's, being scribed by Helen Schucman, with the assistance of William Thetford, of teachings given directly to them by the personage we know as Jesus. It is published by the Foundation For Inner Peace and is being studied by multitudes of people in every corner of the planet, having been translated into numerous languages. The Course consists mainly of the Text, the Workbook for Students and the Manual for Teachers.

These four volumes of posters I have put together with quotes from the Text are meant to be an introduction to The Course for those who do not know of it, and a study tool for those who do. Volume One has ten quotes, one each from the Introduction to Chapter 9. Volume Two covers chapters 10 –19. Volume Three includes quotes from chapters 20 – 29. Volume Four has one quote from Chapter 30, and nine quotes from Chapter 31, the last chapter of the Text, with the last two quotes coming from the 8th and final section of Chapter 31.

I quote the Introduction to The Course:
This is a course in miracles. It is a required course. Only the time you take it is voluntary. Free will does not mean that you can establish the curriculum. It means only that you can elect what you want to take at a given time. The course does not aim at teaching the meaning of love, for that is beyond what can be taught. It does aim, however, at removing the blocks to the awareness of love's presence, which is your natural inheritance. The opposite of love is fear, but what is all-encompassing can have no opposite.
This course can therefore be summed up very simply in this way:
Nothing real can be threatened.
Nothing unreal exists.
Herein lies the peace of God.

For those of us who study The Course, it is life transforming. I have seen the effects of its teachings work wonders in the lives of many who have actively worked at incorporating the lessons of The Course into their lives, and I encourage you to give it a try. Not everybody is ready to receive it, but if you have this book in your hand, I am supposing that you are ready. For some, it is easier to read the Manual for Teachers first, before delving into the Text.

ACIM is NOT a religion. It contains no dogma nor creed. There is no church, priest, minister, rabbi or Iman. You don't need to be baptized or engage in any ritual. You can sit by yourself and study at your own pace, taking from it what you can understand at the time, and go back over it as many times as it takes.

There are many ACIM study groups around the world that meet in person and online if you want to engage in discussion and ask questions. There are videos and audios and other books written about it, but you can be as isolated or involved as you like. A very good resource for additional material is the Foundation For Inner Peace website: https://acim.org

I also caution you not to get sidetracked by the language. Many object to the "Christian" and "male" sounding language. Listen beyond the words. Peoples of all religions, as well as atheists, have benefited from The Course once they understood WHAT was being taught, and not gotten entangled in the words upon first reading. If it is hard for you to accept that Jesus is the author, then concentrate on the substance being taught, not the name of the teacher.

I hope you enjoy and are inspired by the mixture of words and images that I have put together. It was a challenge to find images that I thought would reflect the words, but it was also much fun.

Peace, John Vincent Palozzi

Positive Affirmation Posters

A Course in Miracles 1

*Nothing real
can be threatened.*

Nothing unreal exists.

*Herein lies
the peace of God.*

photo by Harli Marten

Positive
Affirmation
Posters
A Course In Miracles 1

JohnVincentPalozzi@gmail.com

Positive
Affirmation
Posters
A Course In Miracles 1

JohnVincentPalozzi@gmail.com

I am here only
to be truly helpful.
I am here to represent
Him Who sent me.
I do not have to worry
about what to say or
what to do, because
He Who sent me
will direct me.
I am content to be wherever
He wishes, knowing He
goes there with me.
I will be healed as I let
Him teach me to heal.

photo by Jared Rice

Positive
Affirmation
Posters
A Course In Miracles 1

JohnVincentPalozzi@gmail.com

Positive
Affirmation
Posters
A Course In Miracles 1

JohnVincentPalozzi@gmail.com

The Kingdom of Heaven *is* you.

What else **but** you did the Creator create, and what else **but** you is His Kingdom?

Positive
Affirmation
Posters
A Course In Miracles 1

JohnVincentPalozzi@gmail.com

I must have decided wrongly,
because I am not at peace.
I made the decision myself,
but I can also decide otherwise.
I want to decide otherwise,
because I want to be at peace.
I do not feel guilty,
because the Holy Spirit will undo
all the consequences of my wrong decision
if I will let Him.
I choose to let Him,
by allowing Him to decide for God for me.

photo by Adrian Swancar

Positive
Affirmation
Posters
A Course In Miracles 1

JohnVincentPalozzi@gmail.com

photo by Marc-Olivier Jodoin

Positive
Affirmation
Posters
A Course In Miracles 1

JohnVincentPalozzi@gmail.com

The Holy Spirit will always guide you truly, because your joy is His.

This is His Will for everyone because He speaks for the Kingdom of God, which is joy.

Following Him is therefore the easiest thing in the world, and the only thing that is easy, because it is not of the world.

photo by Davide Cantelli

Positive
Affirmation
Posters
A Course In Miracles 1

JohnVincentPalozzi@gmail.com

Positive
Affirmation
Posters
A Course In Miracles 1

JohnVincentPalozzi@gmail.com

Positive
Affirmation
Posters
A Course In Miracles 1

JohnVincentPalozzi@gmail.com

Acknowledgements & Credits

Acknowledgments

The photographs used in these posters were obtained from Unsplash. It is a website that offers photographs for free use donated by professional photographers. Although the License agreement states that credit need not be given, I have credited each photographer on the poster and also in the list below. Please read the Upsplash License, reprinted from their website:

Unsplash License: https://unsplash.com

All photos published on Unsplash can be used for free. You can use them for commercial and noncommercial purposes. You do not need to ask permission from or provide credit to the photographer or Unsplash, although it is appreciated when possible.

More precisely, Unsplash grants you an irrevocable, nonexclusive, worldwide copyright license to download, copy, modify, distribute, perform, and use photos from Unsplash for free, including for commercial purposes, without permission from or attributing the photographer or Unsplash. This license does not include the right to compile photos from Unsplash to replicate a similar or competing service.

Credits: Each poster represents one chapter in ACIM. The numbers reference where you can find the quote in the book (Chapter, Section, Paragraph, Sentence). All quotes are from the TEXT.   www.acourseinmiraclesnow.com
The names to the right are those of the photographers.

01: Introduction.2.2 – 4 (Nothing real)	Harli Marten
02: 1.VII.1.4 & 2.1 (All real pleasure)	DDP
03: 2.V.A.18. (8) (I am here)	Jared Rice
04: 3.VII.5. 5,6 (Your creation)	Ashim D'Silva
05: 4.III.1. 1. 4,5 (The Kingdom)	Will van Wingerden
06: 5.VII. 7 – 11(I must have)	Adrian Swancar
07: 6.V.A.5. 13, 6.V.B.7. 5, 6.V.C.2. 8 (To have)	Marc-Olivier Jodoin
08: 7.XI.1.1-3. (The Holy Spirit will)	Davide Cantelli
09: 8.VIII.9. 1&2 (The Holy Spirit teaches)	Kylo
10: 9.VI.3. 5 & 6 (God has)	Noorulabdeen Ahmad

This is the fifth in a series

Other editions include Affirmations from

Law of Attraction

Quakerism

Made in the USA
Middletown, DE
10 October 2025

19083218R00024